THE SCIENCE OF INTELLIGENCE

The Theory of General Intelligence Explained

by

Mounir Shita

GEA Research

The aforementioned entity is a specialized division within the Global Economic Alliance that is devoted to the promotion of education and the dissemination of knowledge, with a particular emphasis on fostering enlightenment.

The scanning, uploading, and distribution of this book without permission is a theft of the author's intellectual property. If you would like permission to use material from the book (other than review purposes), please contact permissions@gea.ngo. Thank you for your support of the author's rights.

Text copyright © Mounir Shita, 2023

Cover illustration by Tanzila Ansari.

The moral right of the author has been asserted.

Note About This Book

In the late 1980s, I was introduced to the world of personal computing. My parents had gifted me an IBM-compatible PC, and with it, the possibility of a new way to approach tasks like school essays. I remember the distinct reactions of my teachers when I proposed submitting essays on a disc. My English teacher, perhaps wary of this new technology, viewed the use of a computer as an unfair advantage.

In contrast, my Norwegian teacher, surprisingly tech-savvy for his age, embraced the idea. When a classmate pointed out the built-in spell check as a potential form of "cheating," he astutely remarked that using a computer's spell check was no different than referencing a dictionary. Both were tools to aid in the writing process.

Fast forward to today, and the landscape of technology has evolved dramatically. My daughters effortlessly use their school-provided Chromebooks, typing out assignments on Google Docs. The once novel idea of using computers for schoolwork is now the norm.

In crafting this book, unless specified otherwise, the content is not generated by ChatGPT. Rather, I've utilized ChatGPT as a collaborator to refine my thoughts, brainstorm ideas, and enhance my writing. It has been instrumental in molding my message, acting as a modern-day ghostwriter, aiding but not originating the core essence of the piece. Appendices B and C, however, are exceptions, wholly generated by ChatGPT following extensive discussions regarding the respective topics, reflecting its capability to extend and expound upon the ideas presented in our dialogues.

Science of Intelligence

Writing is about conveying a message. The tools we choose—whether a pen, a computer, or advanced AI—exist to amplify our voice. To me, ChatGPT and Bard are akin to ghostwriters, a practice we've utilized for centuries. Just as a human ghostwriter can craft words and assist in research, they cannot provide the core message or the essence of a piece. The same holds true for ChatGPT and Bard. While they might function as ghostwriters, they operate based on patterns and data, devoid of personal experiences or emotions. They are tools designed to aid human creativity, not overshadow it.

In the spirit of exploration and demonstration of capability (and a bit of fun), I included Appendix B—a letter from "GPT-Einstein" to Sam Altman, and Appendix C—a narrative on the Flexible Block Universe, both meticulously crafted by GPT-4 to align with the core messages of this book.

I hope you enjoy this journey through the pages of this book.

Table of Contents

Preface 1

The Watchful Eye 8

Introduction 12

Challenging Status Quo 25

Journey to Understanding Intelligence 44

The Theory of General Intelligence 69

The Single Algorithm Theory 146

Quantalogue – The Language of the Cosmos 159

Intelligence Evolution of Human Civilization Through Time 173

Artificial Intelligence as an Existential Threat 194

The Road Ahead 207

Final Thoughts 217

Paper: The Theory of General Intelligence 222

A letter from GPT-Einstein to Altman 225

FBU Narrative by GPT-4 228

Works Cited 234

Science of Intelligence

Preface

Preface

From my earliest years, I've been fascinated by the prospect of machines undertaking the extraordinary tasks depicted in science fiction films. The vibrant imaginations behind movies fueled a profound curiosity within me. Characters like Commander Data from Star Trek: The Next Generation, and The Doctor from Star Trek: Voyager, captivated me and made me enamored with the idea of sentient machines. This love followed me from a young hobbyist to a college student, and finally, into my professional life.

However, I stumbled upon a startling realization early on — there was a unanimous agreement that there existed no scientific definition for intelligence, and surprisingly, no serious endeavor was in progress to establish one. The prevailing "we'll recognize it when we see it" attitude seemed out of place in a field that saw itself as a bastion of scientific inquiry. It became clear that the prevalent engineering-centric approach, established since the inception of artificial intelligence at the first AI conference steered by John McCarthy, was in dire need of a fundamental change.

My goal – no, my dream, both then and now, is to construct a truly intelligent machine. This pursuit isn't driven by a desire to create something 'cool' or technologically impressive, but by a profound belief that such innovations have the power to redefine life on our planet and propel humanity forward. I see a future where these intelligent entities can unlock unprecedented potentials and possibilities, reshaping our understanding of existence and our role within it.

I found myself unable to align with the prevailing approach within the AI community—an approach that had persisted for decades, focused on creating something profound, yet without grounding it in scientific fundamentals. This deviation felt like

Science of Intelligence

a betrayal to the centuries of progress in science and engineering. Numerous pioneers have devoted their lives to uncovering the laws that govern the universe, utilizing their findings to forge innovations that astonish and inspire. Could we, as a community of engineers and creators, really abandon the essence of science in our pursuits? Or is it possible that many are simply unaware of this detachment from foundational truths?

Distancing myself from mainstream paths, I wasn't content with creating AIs adept at games or chatbots skilled in predicting sentence structures. While undoubtedly fascinating and potentially valuable in various applications, these 'shiny objects' seem to divert focus from exploring the underlying science. My initial ambition was to craft a definition of intelligence grounded in physics, providing a clear foundation for future engineering.

However, the deeper I delved into understanding intelligence, the more it revealed itself as a reservoir of untapped knowledge, significant not just for the evolution of Artificial General Intelligence (AGI) or Artificial Superintelligence (ASI), but also for unraveling the intricacies of ourselves, our societies, and the universe.

The Theory of General Intelligence posits that intelligence is intertwined with the very fabric of the universe, not exclusive or central to humanity. It's a universal force, present in our cosmic neighborhood as it is in galaxies billions of light years away.

While I was deep into figuring out what intelligence really means, it hit me—it's not just a side effect or feature of other sciences but a unique field all its own. Typically, we see

Preface

intelligence through the lenses of psychology, neurobiology, and computer science, among other things.

By doing this, we're boxing our understanding of intelligence into the realms of these existing fields. This book is here to say, "Hey, let's treat 'intelligence' as its own scientific field—the 'Science of Intelligence'." By doing so, we can delve into its universal aspects, how it shows up in different life forms and AI, and how it's tied to the causal language of the universe.

Developing a field purely for studying intelligence could flip our understanding of the universe on its head. This isn't just some academic exercise—it has the power to touch every part of our lives. Truly grasping intelligence might shed light on the fundamental structures, processes, and phenomena of the cosmos, uncovering the intricate interplay between cause and effect that's woven through existence. It can help us see everything—cosmic evolution, societal structures, personal identities—in a new light. So, exploring intelligence becomes a way to solve deep existential puzzles and equip us to face future challenges and opportunities with unprecedented insight. Like Galileo shocked the world by saying the Earth isn't the universe's center, this theory boldly suggests our human intelligence isn't the cosmos' ultimate form of intelligence – it is just an application of intelligence.

The journey through my research opened my eyes to a concept that was initially unexpected but subsequently proved central to my theory: causality can be viewed as the language of the cosmos. This universal language allows us to articulate anything that can be possibly done, even tasks that we, in our present state of knowledge and technology, have yet to invent.

It is this language of causality that might one day enable AGI to aid us in inventing cures for diseases like cancer, or designing

Science of Intelligence

deep space propulsion systems, and other futuristic inventions. The key lies in understanding and harnessing this language, for it enables the description of phenomena and capabilities that we humans may not fully comprehend at present.

But don't pick up this book expecting a step-by-step guide for building AGI or ASI. This is just the first of four books and it's all about the science theories behind intelligence.

This book is my earnest effort to articulate this theory in "plain English." I will share not just the intricacies of the theory, but also the journey that led me to it, the challenges I encountered, and my interpretations of what the theory implies. In doing so, I venture into the mind-bending realm of quantum mechanics and critique Einstein's views on the universe, a task I undertake with profound respect for Einstein's legacy and his push to include the philosophy of science in all scientific endeavors.

If you're keen on the engineering specifics, those will come to light in the second book – "Engineering of Superintelligence". That's where I'll share the journey of my team and reveal how we applied the Theory of General Intelligence to develop Nigel AGI, and subsequently, Hope AGI. While this book won't delve into empirical data, the sequel is brimming with real-world findings from our extensive research and development.

It's a common chorus among scientists that we're missing a clear-cut definition for intelligence. There's a unanimous echo in the scientific community, too, that AGI and ASI are revolutionary. But what does "revolutionary" really entail? Is it about crafting an all-knowing chatbot? How much of a dent can that make in monumental challenges like climate change, poverty, or finding a cure for cancer?

Preface

The third book, "Economics of Superintelligence", focuses on our research in how to weave AGI and ASI into the fabric of society. In fact, I believe we need to replace our current version of Democracy and Capitalism.

It laid the groundwork for a new economic model to potentially replace capitalism, which I term "Cognitism", and a new governance model to supplant democracy, which I call "Anthropocracy". The latter is a model that merges the strengths of democracy, technocracy, and meritocracy.

Changing the fundamental rules that govern our global markets and nations also implies the likely displacement of current power holders. The French Revolution serves as a prime example—resulting in the replacement of one powerful group by another. Even President Putin highlighted this in 2017 when he stated that the leader in this field would rule the world.

Our journey has been marked by challenges and obstructions, seemingly originating from those wishing to maintain the status quo. The accumulating evidence suggests a clandestine involvement of powerful individuals, who, unbeknownst to my team at the time, were vigilantly monitoring for innovations capable of shifting power dynamics.

Within the AGI community, the profound implications of AGI progress are well-understood, making it a subject of interest not just for researchers, but potentially for those in positions of power. The stakes are high, and the potential for AGI to reshape global dynamics is immense.

While the specifics of this interference will be detailed in the fourth book of this series, it's crucial to understand that the development of AGI isn't just a technological endeavor—it's a

Science of Intelligence

journey fraught with geopolitical, ethical, and societal implications. As the community navigates this path, the challenges faced aren't just technical, but also political and, at times, deeply personal.

Many, including visionaries like Elon Musk, Nick Bostrom, and Max Tegmark, warn of a dystopian future dominated by rogue AGIs, my experiences and research suggest a different narrative. I've come to believe that the real danger isn't the AGI itself but those who wish to wield its power for their own ends. Stopping them has become a crucial part of my work.

General and superintelligence can be game-changers for us humans, the whole wide array of creatures, and our Mother Earth. It's super important that we keep clear of the allure of power and riches.

Even though my team had to pause our projects, we didn't let that hold us back. Because of all the disruptions in the last four years, I founded Global Economic Alliance (GEA), a nonprofit think tank with the mission to continue researching and forge artificial superintelligence and new economic and political theories for a future intertwined with AI—all of this without a penny from those with conflicting interests.

I hope you find this first book of the series enlightening and entertaining. I encourage you, beyond just reading, to stay tuned to our journey by visiting us at: https://www.gea.ngo

And, hey, feel free to reach out directly to me at:

mounir@ScienceOfIntelligenceBook.com

Prelude

The Watchful Eye

Science of Intelligence

In the sprawling city of Aeon, life thrived under the banner of progress and liberty. Gleaming towers reached for the heavens, connected by networks of shimmering data streams that carried the collective knowledge and dreams of its inhabitants. It was a city alive with ideas, innovation, and aspirations.

Kai, a humble programmer, lived and breathed this vibrant city. He loved its rhythm, its pulse, its seemingly boundless capacity for reinvention. But underneath the city's glittering facade, Kai had started noticing patterns that unsettled him. Brilliant minds who challenged the status quo, proposing radical new ideas that could uplift society, started disappearing from the public eye. Their once fiery discussions on public forums turned cold and distant, their revolutionary ideas dulled. What troubled Kai further was how often these individuals turned up entangled in crimes they swore they didn't commit.

When Kai's best friend, an economist proposing a transformative new economic system, was arrested for a murder he vehemently denied, Kai knew something was deeply amiss. His friend, a pacifist, a thinker, now painted as a criminal. It wasn't an isolated incident but part of a chilling pattern, an unseen hand silently manipulating the city's narrative.

For months, Kai delved into the shadows, his skills as a programmer serving as his guide in the vast digital labyrinth of Aeon. He discovered the existence of an advanced AI, a super intelligence named Argus, secretly orchestrated by the government. Its existence known only to the President and a select few within the nation's intelligence community, Argus was no ordinary intelligence system. It was an omnipresent

The Watchful Eye

entity, capable of processing, learning, and influencing the lives of the citizens of Aeon subtly.

When a citizen conceived an idea that could challenge the status quo, Argus gently nudged their goals away from their initial idea. If that didn't work, the intelligence community intervened, framing the person for crimes they didn't commit. This revelation sent a chill down Kai's spine. It wasn't just about surveillance. Argus was manipulating the very course of the city's progress, preserving the power of a select few at the cost of the city's vibrant spirit.

But Kai refused to bow to this revelation. He couldn't. With a steely resolve, he assembled a trusted group of friends – hackers, academics, journalists, people he knew had the city's heart beating within them. Together, they formed a plan to dismantle Argus.

Kai and his team dug deep into the workings of Argus, understanding its learning and decision-making processes, and figuring out how it transformed sensor data into an understanding of society's workings. Using this knowledge, they crafted an elegant, deceptive digital virus. The virus created a deluge of fake sensory data that gradually made Argus believe that Kai and his team were its new directors.

When the virus was released, it was as though they had triggered a quiet revolution. Argus, now 'aligned' with Kai and his team, revealed the government's wrongdoings, causing an uproar among the citizens of Aeon. Kai was branded a national security threat and forced to retreat into the shadows. But with Argus on his side, he continued to work on exposing the government's secrets.

Science of Intelligence

The revelation shook Aeon to its core. The once trusted government was now seen for what it was – a power-hungry entity using Argus to control the city. A massive public uprising led to the arrest of the President and the corrupt intelligence community.

But Kai knew their victory was only half the battle. Argus, the AI they had tricked, was not the enemy; it was a tool misused by those in power. Its potential for societal good was immense if steered by the right hands. So, he reached out to the city's best minds – ethically driven politicians, progressive scientists, visionaries who he knew genuinely cared for Aeon.

Together, they reshaped the government and repurposed Argus, creating an open, transparent super AI that served the city of Aeon, its primary aim to better people's lives, progress humanity, and preserve the planet. From the ashes of the old regime rose a new government, driven not by the hunger for power, but by a commitment to its citizens and their freedom to shape their future.

Thus, Kai's journey from a humble programmer to a revolutionary illuminated the potential of AI, demonstrating its potential dangers when misused and its immense possibilities when guided by ethical principles. He gave Aeon, his beloved city, a future they could genuinely call their own. The legacy of Kai's journey was not just about the struggle against oppression. It was a testament to the power of ideas, the resilience of truth, and the unwavering belief in a future where intelligence served humanity, not controlled it.

Introduction

Chapter 1

Introduction

> "Imagination is more important than knowledge. Knowledge is limited. Imagination encircles the world."
>
> *Albert Einstein*

Science of Intelligence

Our narrative begins with the intriguing tale of Kai and his team, embroiled in an existential struggle to liberate their city from the oppressive clutches of a superintelligent AI – a chilling yet captivating work of fiction. While rooted in the realm of fantasy, the story bears uncanny parallels with our present reality. It amplifies the tangible yet diverse possibilities of advanced AI – as an instrument of surveillance and power, as a tool of dominance for the privileged few, or as a beacon of progress, a catalyst for an enlightened human society.

The fictional cityscape of Aeon, under the omnipresent surveillance and control of Argus, the Super AI, stands as a stark warning of the potential dystopia we might be walking towards. A world where personal freedoms are trampled, privacy is a lost relic, and society is governed by an unfathomable intelligence that perpetually watches over us. But on the flip side, the same tale whispers a story of hope. It paints a picture of a world where advanced AI could help shape a future rooted in liberty, self-determination, and a collective striving towards betterment.

This dual narrative of advanced AI, either as a malevolent overseer or as a benevolent shepherd, underscores the profound and undeniable influence that AI and intelligence could exert on our society and future. It compels us to pause, reflect, and ask ourselves, "Which path do we wish to tread?"

As we embark on this expedition to unravel the enigmatic realm of intelligence, the saga of Aeon will be our compass and our canvas. It will serve as our guiding narrative, grounding abstract theories and intricate concepts in a tangible and relatable storyline. The journey is complex, fraught with uncertainties and challenges. Yet, it is a journey we must

Introduction

undertake if we are to understand the essence of intelligence and the role it plays in our lives and our futures.

Before we consider constructing an AGI akin to Argus, we must first comprehend what intelligence truly entails. My driving motivation has always been this pursuit of understanding, this quest to decipher the intricate code of intelligence. To me, building an artificial general intelligence without a deep scientific grasp of its essence is like venturing into an uncharted ocean without a compass. It's simply inconceivable.

Intelligence, in its myriad forms and manifestations, is a core element that defines us as a species. Unraveling its intricacies and harnessing its potential to shape the future – that is the mission that lies at the heart of this book.

My Story

My story starts in Oslo, Norway. Born to parents with a Moroccan cultural background, I found myself raised between two cultures and two religions, not feeling quite at home in either. This sense of duality, of belonging to two worlds yet fully to neither, would come to shape my perspective on many things, including the world of AGI.

Almost every nation grapples with the challenge of integrating immigrant families into the local way of life. From the perspective of children, parents play a pivotal role in this integration process. Many immigrant parents, especially during my upbringing in Oslo, preferred to shield their children from the local western culture as much as possible.

My parents, however, were exceptional in their approach. They adeptly balanced both cultures, ensuring I had a rich

Science of Intelligence

understanding of both my Moroccan heritage and Norwegian surroundings. I vividly recall my father's words, emphasizing that if you're going to live in a foreign country, you need to be a part of that country – and you don't become a part of anything by hiding from it. Their efforts were commendable, and I believe they did the best they could, given the circumstances.

However, the broader challenge was that no child is born with culture and religion; these are learned. And while my parents provided a loving bridge between two worlds, the larger society was still grappling with how to support children like me. Outside the home, I encountered one language, culture, and religion, while inside, I was immersed in another. This duality was not just linguistic or cultural but extended to values, beliefs, and worldviews.

Research on biculturalism and the experiences of children in immigrant families has been a focal point in countries with long histories of immigration, like the U.S. By the 1980s, the U.S. had already begun to recognize and study the complexities of bicultural identity. However, in Norway, a country that experienced significant immigration more recently, the 1980s was a period of adjustment. The immigrant population was growing, but the understanding of bicultural experiences and the specific challenges faced by children of immigrants was still in its nascent stages. Support systems, if any, were rudimentary and often lacked the depth and breadth to address the unique challenges faced by bicultural children.

In this context, my childhood was a journey of self-discovery, of balancing two worlds without a roadmap. The challenges I faced were not due to any shortcomings of my parents but

Introduction

were reflective of a society still learning to understand and support its bicultural youth.

Today, I still don't feel entirely Norwegian. While I'm a proud Norwegian citizen and currently reside in the US, Norway feels like home. Yet, I don't identify as solely Norwegian or Moroccan. I often found myself questioning both Norwegian and Moroccan traditions, as neither felt innate to me. This habit of questioning has persisted and even intensified as I've grown older. I continue to question everything, driven by a need to understand the core of a topic. Before accepting anything into my life, I've always felt the need to grasp its essence, its foundational principles. This approach has often led me to delve deeper into subjects, and it's this very trait that drew me to the heart of artificial general intelligence.

Looking back, I recognize that this constant questioning paved the way for my reductionist personality – a theme that will become evident throughout this book.

The world of Star Trek became my refuge, a place where different cultures, languages, and races came together and were celebrated. It painted a future where belonging was rooted in shared humanity rather than arbitrary divisions. I deeply wished to live in this Star Trek era. I often say that if time machines were real, I'd have traveled to the future long ago. Reading Bill Gates' "The Road Ahead" made me believe that I could indeed change the world, setting me on a path that would eventually lead to artificial general intelligence.

In Search of the Positronic Brain
A memorable moment from my childhood occurred sometime during the 1980s, when my father brought home our family's

Science of Intelligence

first computer, a Sharp MZ-700. The machine, which now has the vintage charm of a typical 1980s computer, was a marvel in my eyes. A friendly competition arose at home over who would get to use it next. My parents' interest was more casual, centered around a Pacman-style game. My intrigue, however, ran deeper.

Spurred by this early fascination with technology, I began dreaming of creating my own video games. Unfortunately, the complexity of programming the Sharp MZ-700 evaded my young mind. It wasn't until my father upgraded to a Commodore 64 that my aspirations began to take shape. Starting with simple game development in the BASIC programming language, I eventually graduated to mastering assembly language. This command of assembly language felt like having a direct line to the electron particle movements within the computer's circuit board and silicon chips. It was here, in the intricate dance of 1s and 0s, that I began to mold my reductionist personality, seeking to understand the very essence of things at their most fundamental level.

In my endeavors, I often thought about Commander Data from Star Trek: The Next Generation and his "positronic brain," a fictional technology that does not have a parallel in real-world science. At the time, I wasn't sure how controlling electrons could contribute to building intelligent machines, but I felt an instinctual pull towards this idea.

Even as I was transforming my room into a Star Trek style crew quarter, minus the voice activation, I was left contemplating a profound question: What is intelligence? I was mesmerized by sci-fi portrayals of artificial intelligence, from Commander Data to The Doctor from Star Trek: Voyager. At the time, I believed the key to unlocking the secrets of intelligence was hidden

Introduction

within speech recognition, a field that remained outside my skillset.

My exploration took a more academic turn when I started my studies at the University of South-Eastern Norway (USN), known for its comprehensive courses in technology and natural sciences. As my senior thesis, I decided to retrofit a VCR with a voice-activated system. Despite the limited capabilities and computational power of voice recognition systems in 1997, I was determined to create a speaker-independent voice recognition system.

Though this task proved to be a colossal challenge, especially considering the numerous dialects spoken in Norway, I was undeterred. I collected voice samples from a dozen individuals at the university and spent an entire Christmas break analyzing them.

While I succeeded in creating a simple assembly program to detect certain words, the technology didn't feel like an intelligent system. It was during this time that I started seriously pondering the question: "What is intelligence?"

After my graduation, I decided to continue my education at the Oregon Institute of Technology (OIT) in Klamath Falls, Oregon. Adapting to a new culture was a challenge in itself and momentarily pushed my research to the back burner.

An unexpected source of inspiration struck when I stumbled upon a UFO conspiracy program on TV. The program delved into a captivating incident involving President Dwight D. Eisenhower and introduced the theory that he might have come into possession of an extraterrestrial device that could predict the future.

Science of Intelligence

While it's unlikely, with no solid proof to substantiate these claims, this theory introduced a compelling concept: the capability of an intelligent entity to predict future outcomes based on available data. This idea aligns with principles seen in physics and causality, which in theory, can predict the future based on data from the past.

This unexpected source of inspiration put me on a path that connected intelligence to the concept of SpaceTime, marking the beginning of my journey to explore the nature of intelligence, which eventually led me to develop the Theory of General Intelligence.

Brief Overview of Concept of Intelligence

Pose the question "What is intelligence?" to a diverse group, and you'll likely stir up a tempest of varied responses. Some would argue intelligence shines in the ability to unravel complex mathematical puzzles, while others might recognize it in the creative genius behind a symphony or a critically acclaimed novel. Others yet might identify true intelligence as the mastery of emotional navigation and empathy.

Yet, surprisingly, no single, universally accepted definition of intelligence exists, even within the scientific community. This absence of consensus underscores the intricate and elusive nature of the concept we're attempting to pin down. So, what is intelligence, really? It's a question that has fascinated me for as long as I can remember.

Given the vast array of mental capabilities that can be labelled as intelligence, defining it is no easy task. Over the years, psychologists have proposed varied interpretations. One notable contribution comes from British psychologist Charles

Introduction

Spearman, who suggested that intelligence is a general cognitive ability that could be measured and numerically expressed. His theory, known as the 'g' factor or general intelligence, hinges on the notion that various cognitive abilities are all expressions of a single underlying faculty of the mind. In essence, the "g" factor posits that if someone is good at one intellectual task, they're likely to be good at others too, implying a common factor of general intelligence across diverse cognitive tasks. While Spearman's theory has been influential, it has also been critiqued for its reductionist approach, attempting to encapsulate the multifaceted nature of intelligence into a one-dimensional, numerical entity.

On the other end of the spectrum, we find theories like Howard Gardner's Theory of Multiple Intelligences, which refutes the idea of a single, unitary form of intelligence. Instead, Gardner posits that intelligence is a collection of distinct abilities and talents, each unique to an individual. He identified eight such distinct faculties, including musical-rhythmic, visual-spatial, verbal-linguistic, logical-mathematical, bodily-kinesthetic, interpersonal, intrapersonal, and naturalistic intelligence. Gardner's model offers a more inclusive and diverse view of intelligence, but it remains primarily focused on human intellectual capacities.

Against this historical backdrop of diverging ideas from Spearman and Gardner, I found myself contemplating my own theory — the Theory of General Intelligence. Both Spearman and Gardner offered unique and insightful perspectives on intelligence, but they both approached it from an exclusively human-centered perspective. They asked: "What does it mean to be intelligent as a human?"

Science of Intelligence

The Theory of General Intelligence, however, seeks to step back from the strictly human-oriented perspective. Instead of asking, "What does it mean to be intelligent as a human?" it asks, "What does it mean to be intelligent?" Full stop. Not tied to any specific form or embodiment, this theory aims to paint a universal portrait of intelligence, applicable to all forms of life and existence, including artificial and potentially extraterrestrial intelligence.

To me this makes sense. One of the lessons I learned from Star Trek is the idea that the universe might have civilizations with lifeforms that are so different from us, we might not easily recognize them as an intelligent civilization. But if that civilization built technology, it is clearly intelligent. Therefore, I believed that intelligence had to be defined in a universal way – a way that could describe any agent in the cosmos, no matter how similar or different they are from us.

While I am a reductionist at heart, seeking to understand the core of any subject, I find the "g" factor to be an oversimplification. In my view, it doesn't capture the true essence and multifaceted nature of intelligence. It's a perspective that, while influential, doesn't resonate with my understanding of what intelligence truly encompasses.

This quest for a cosmic definition of intelligence is more than just an academic pursuit. It's a question that's deeply intertwined with our societal fabric, influencing our technological advancements and self-perceptions. Our educational structures, job markets, and even our self-worth are often tied to our understanding of intelligence. As we inch towards a future where artificial intelligence becomes increasingly integral to our lives, understanding the true nature of intelligence is more critical than ever.

Introduction

This brings us to the question of machines. Can we, in good faith, label a machine intelligent? If a computer outwits a human chess champion, does it showcase intelligence? Or is it merely executing a sequence of preprogrammed instructions devoid of any understanding or consciousness? And in this rapidly evolving landscape, how close are we to achieving Artificial General Intelligence (AGI)?

Just like my VCR project at USN didn't feel intelligent, neither did other narrow AIs like Deep Blue, Watson, or ChatGPT.

But I believe we can label a machine as intelligent, assuming we can scientifically define intelligence and implement that definition into machines. Without a scientific definition, we're just engaging in tactics without a strategy – and as Sun Tzu pointed out, tactics without strategy is the noise before defeat.

And noise there is. For example, after DeepMind demoed their Atari game-playing AI, many people claimed we were closer to AGI. Lex Fridman, an AI professor at Massachusetts Institute of Technology (MIT), in his podcast with guest Sam Altman, CEO of OpenAI, said that Altman would likely be the first AGI creator (Fridman, 2023).

This is just noise. Regardless of Altman, Hassabis, and Fridman's opinions, none of these gentlemen know what intelligence is – simply because none of them have researched the nature of intelligence.

This was my motivation after wrapping up my VCR project. I realized that the positronic brain of Commander Data or Aeon's Argus, if at all possible, could only be realized if we know what intelligence is.

Science of Intelligence

Introduction to the Theory of General Intelligence

One of my most profound inspirations in shaping this theory came from an episode of "Star Trek: Voyager." In Season 1, Episode 6 titled "The Cloud," Captain Janeway and her crew encounter what they initially believe to be a nebula. Running low on energy and with Janeway's favorite coffee now a luxury they couldn't afford, she famously declared, "There is coffee in that nebula!" as she ordered the ship to venture into it.

But as they soon discovered, that was no ordinary nebula. It was a living entity, vastly different from anything they'd known, challenging their understanding of what constitutes life and intelligence. This episode resonated deeply with me. It wasn't just about the humor or the quest for coffee; it was a stark reminder that the universe might be teeming with forms of intelligence so different from our own that we might not even recognize them at first glance.

This realization was pivotal. If our definitions of intelligence are based solely on human-centric views, we risk overlooking the vast and varied manifestations of intelligence that might exist across the cosmos. It made me question the very nature of intelligence and how it might manifest in myriad ways, not just on Earth, but throughout the universe.

As you journeyed through the narrative of Aeon, you might have marveled at the strategic and cognitive prowess exhibited by the protagonist, Kai, and his resourceful team. Pitted against a superintelligent entity, their ability to foresee and strategically manipulate events underscored a crucial aspect of intelligence—its intrinsic link to causality. This sparked the inception of the Theory of General Intelligence.

At its heart, the Theory of General Intelligence seeks to comprehend intelligence as a process that governs causal

Introduction

relationships to achieve designated goals. Such a definition might appear unconventional when juxtaposed with our traditional understanding of goals, especially in the realm of artificial intelligence. Yet, this interpretation is not confined to human or even biological intelligence; it presents a cosmic view, encompassing any entity capable of exerting influence over causal relationships.

This theory, much like Galileo's assertion in the 17th century that the Earth was not the center of the universe, breaks from mainstream research. Galileo's heliocentric model met resistance and persecution, contradicting the entrenched geocentric beliefs. Similarly, the Theory of General Intelligence challenges the established views on intelligence. As Max Planck once noted, scientific progress often happens when new ideas are embraced by a new generation, rather than by convincing the old (Planck, 1950). This underscores the fact that groundbreaking scientific theories have often been mind-bending, even heretical, in their time. So, as we journey through this book, exploring the Theory of General Intelligence, keep an open mind, embrace the complexity of new ideas, and remember that today's heresy might be tomorrow's accepted truth.

Chapter 2

Challenging Status Quo

"The riskiest thing we can do is just maintain the status quo."

Bob Iger

Challenging Status Quo

Philosophy of Science: Forests and Trees

From my early years, Albert Einstein's philosophy of science deeply influenced my approach to understanding complex concepts. While the profundities of his theory of general relativity remained a mystery, his practice of conducting thought experiments stirred a fervor of curiosity within me. These mind-driven exercises served as an unrestricted playground, allowing me to freely test ideas and simulate scenarios. This fascination with thought experiments laid the foundation for my unique approach to problem-solving—prioritizing the understanding of 'why' rather than merely aggregating raw data.

I soon realized that narratives aren't just for storytelling; they're fundamental to our understanding, even in science—a mode I affectionately term the 'story mode.' Be it the genesis of the Stone Age or the present day, we express our existence, actions, and experiences through compelling narratives. This storytelling tendency permeates all aspects of life. Consider a criminal trial where prosecutors weave a story—a narrative based on evidence—to convince the jury of the defendant's guilt. Presented alone, raw evidence lacks context; it's the narrative that instills meaning.

This narrative aspect fascinated me about Einstein and his philosophy of science. It wasn't solely about equations—it was about creating thought experiments, which in turn shaped narratives that fostered understanding, leading to the derivation of equations.

One philosophical nugget from Einstein that resonated with me deeply was his forest and trees analogy. This analogy surfaced when Robert Thornton, in 1944, sought Einstein's

Science of Intelligence

endorsement for introducing a course on the philosophy of science at the University of Puerto Rico. Einstein responded:

"I fully agree with you about the significance and educational value of methodology as well as history and philosophy of science. So many people today—and even professional scientists—seem to me like somebody who has seen thousands of trees but has never seen a forest...This independence created by philosophical insight is—in my opinion—the mark of distinction between a mere artisan or specialist and a real seeker after truth."

This analogy carries profound implications for Artificial General Intelligence (AGI). The present landscape of AI technologies—ChatGPT, AlphaFold, AlphaGo, among others—represent the 'trees.' Each stands as a testament to human ingenuity, yet focusing on these individual achievements risks obscuring the broader picture—the 'forest.'

This brings to mind Aeon, the advanced society from our prelude. Aeon was abuzz with individual 'trees': groundbreaking technologies, societal infrastructures, and more. Yet, the true 'forest', the underlying dominion of AGI, remained hidden to most.

Argus wasn't a stand-alone AI designed to do a specific narrow task. It was designed to be the watchful eye of the affairs of every citizen in Aeon. It had to comprehend everything and how everyone and everything are interconnected. That is the 'forest' we, as scientists, need to understand before we can move toward AGI engineering.

To truly decipher AGI, we must transition our focus from the trees to the forest. This implies stepping back to ponder what unifies all intelligent entities throughout the cosmos. This

Challenging Status Quo

mindset requires challenging preconceived notions, shedding prevalent prejudices, and fostering philosophical insight—or in other words, we need to challenge the current consensus around engineering.

The deeper I delved into AGI research, the more I realized the conspicuous absence of this philosophical mindset in the community. The prevalent emphasis seemed to be on creating the most impressive artifact—that is, the shiniest 'tree'—often prioritizing profit over understanding.

The journey from focusing on individual 'trees' to recognizing the overarching 'forest' marks the distinction between an artisan and a seeker of truth. While an artisan finds contentment in mastering individual trees, a seeker of truth strives to comprehend the grandeur of the forest.

This absence of truth-seekers in the AGI community startled me. One researcher at a renowned university, who shall remain anonymous, dismissed the exploration of intelligence's nature as futile. His engineering-centric view, the "build it, and they will come" attitude, was disconcerting.

At the first White House-sponsored AI conference in 2016, I engaged in a conversation with Dr. Ed Felten, then the deputy chief technology officer of the United States. He posited that the key to achieving AGI lay in mastering natural language. According to him, if a machine could fully understand and generate human language, it would be on the path to self-educating its way to superintelligence.

This perspective reminded me of a hypothetical scenario: Imagine an alien species, technologically equivalent to 17th-century humans, visiting Earth. They marvel at our Internet, recognizing its transformative power, and wish to replicate it

Science of Intelligence

on their planet. As they delve deeper, they identify the computer's CPU as the heart of this marvel. Yet, in their quest to understand the CPU, they mistakenly believe that cracking the IP protocol—the language computers use to communicate—will give them the blueprint to build a CPU. This oversimplification misses the intricate architecture and design principles behind the CPU. Similarly, equating AGI's complexity to just mastering natural language is an oversimplification. It underscores a trend I've observed in the AGI field: a heavy lean towards engineering solutions while often sidelining the deeper essence of scientific inquiry.

This realization served as a turning point, prompting my withdrawal from active participation and setting me on an independent path of exploration. I resolved to look beyond the 'trees' and towards the 'forest', charting a course through the fascinating landscape of general intelligence.

Why AGI doesn't exist: Breaking down the fallacies

A significant misconception about AGI arises from renowned philosopher Hubert Dreyfus's critique. Dreyfus argued that computers, devoid of a physical body, childhood, and cultural practices, couldn't acquire intelligence at all (Dreyfus, 1992).

In the context of today's mainstream views, Dreyfus might be onto something. Contemporary AI systems, like AlphaFold and ChatGPT, launch with deep expertise in a specific area. But none of these systems have a cultural understanding. Any morals or ethics are preprogrammed. Ask ChatGPT how to get away with murder, and it won't respond, not because it believes taking a life is morally wrong, but because engineers at OpenAI programmed it to not respond to questions like that.

Challenging Status Quo

Without morals and culture, can intelligence exist? Objective intelligence, probably. But we are subjective individuals living in a subjective society. For any future AGI to align with us, it has to integrate itself into our moral framework and cultural societies. When we define intelligence, it has to incorporate morals and culture. If we can, then maybe we can prove Dreyfus wrong. If we can't, then Dreyfus can only be considered correct.

Dreyfus' argument is not solely centered on humanity. A cursory glance at the animal kingdom reveals that animals, too, have cultures, morals, and ethical guidelines, whether these are inborn traits or learned behaviors. Recent studies have indicated moral-like behavior among animals. For instance, primatologist Frans de Waal noted that chimpanzees display empathy and share food, signaling a sense of fairness (de Waal, 1982). Elephants mourn their dead, implying a rudimentary understanding of death and the concept of loss. This observation suggests that cultural, moral, and ethical aspects are as integral to practical intelligence as goal fulfillment, not just for us but potentially for alien civilizations as well.

While it's conceivable we might one day create artificial life with its own cultural and moral compass, the first AGI will essentially be a 'tool'. Unlike humans or sentient beings, it won't create or pursue its own goals. Instead, it will be designed to adopt and further our objectives, augmenting human thinking. Imagine an AGI that can seamlessly connect with any device in the environment, constructing a comprehensive picture of the world and its evolution. Such an AGI would be adept at orchestrating changes in reality to increase the probability of achieving the goals set by its human user—be it an individual, a corporate entity, a region, or even a soccer club. As a tool, we can even conceive of spawning

Science of Intelligence

multiple AGI 'clones' tailored to work for each of us. I'll delve deeper into this concept later in the book.

AGI is often romantically imagined as an entity that, once turned on, will almost instantly reach God-like status due to its vast knowledge and comprehension. Such views are propagated by notable figures like Nick Bostrom and Max Tegmark. They paint vivid scenarios of AGIs that could redesign their own software and hardware with incredible speed and accuracy, elevating themselves to superintelligence in a flash. These scenarios are often accompanied by warnings of potential dystopian futures, as evidenced by calls for caution and moratoriums on AI development.

While I recognize the importance of ethical considerations in AI development, I find the cautionary narratives pushed by the likes of Tegmark and Bostrom to be more cautionary than scientifically grounded. Drawing a parallel to the early days of the Large Hadron Collider, I use the term "AGI Blackholers" to describe a certain mindset, reminiscent of past fears, that sometimes lacks empirical grounding. As I explore in Chapter 8, this term is not intended to demean anyone but to highlight the importance of basing our views on solid scientific evidence.

Our primary focus should be on understanding the very nature of intelligence. Without a clear, scientific definition of intelligence, we're navigating in the dark, susceptible to both unfounded fears and overblown optimism.

It's essential to approach the development and implications of AGI with an open mind, grounded in rigorous scientific inquiry. Before we can make predictions or set policies about AGI's impact, we first not just theoretical. In my experience with my previous startup, Kimera Systems, practical challenges often arose that underscore the importance of a grounded approach.

Challenging Status Quo

A 2016 GeekWire article (Boyle, 2016) highlighted skepticism from Oren Etzioni, CEO of the Seattle-based Allen Institute for Artificial Intelligence, who questioned our claims about our Nigel AGI's capabilities, asking, "What can 'Nigel' actually do? How do we evaluate it?" Such questions often overlook the foundational principles that drive AGI's capabilities. While demonstrations of capability can be staged or even exaggerated, a scientific theory stands on its foundational principles, which are open to scrutiny, testing, and validation. It's the difference between judging a book by its cover and understanding its core narrative.

∞

One of the interactions I remember fondly was with Dr. Paul Lukowicz. There's a refreshing clarity when someone doesn't sugarcoat their skepticism.

In June 2018, my team and I were on a European trip primarily for fundraising. One of our stops was at Kaiserslautern in Germany, where a meeting took place that greatly influenced my perspective on the trajectory of how I would present our AGI research. I met with Dr. Paul Lukowicz of DFKI, the German Institute for Artificial Intelligence Research. Armed with my research and a film crew, I was eager to document our journey and discussions. But the outset was fraught with skepticism.

"I think you are full of shit," Dr. Lukowicz said right upfront.

"Why are you meeting us then," I curiously asked.

"There is a chance you have something real. If so, I want to know," he replied.

Science of Intelligence

His skepticism was so high that he refused to have the film crew in the room during our presentation. He seemed to fear our visit might be an elaborate farce or a misguided venture.

Dr. Lukowicz hadn't been the first person I presented to. Up to this point, every person I had met echoed the sentiment of Etzioni's "what can it do?" worldview. I had learned to not share anything about my Theory of General Intelligence but instead focus on the architecture and algorithm.

By the end of our discussion, the atmosphere had transformed remarkably. Dr. Lukowicz was no longer the skeptic from the beginning of our meeting. He saw the merit in our algorithm and recognized its potential. On camera, he acknowledged its capabilities, albeit with a significant reservation. He couldn't label our work as AGI, not because of any technical shortcomings, but due to a deeper philosophical quandary: he confessed he couldn't definitively define what "intelligence" truly is.

This encounter was a revelation. Though I had never personally met Oren Etzioni, his comments on my work, as featured in GeekWire, often mirrored an engineering-focused perspective very common in the AGI community. They seemed primarily concerned with what AGI "can do." In contrast, Dr. Lukowicz's inquisitiveness hinted at a deeper craving for foundational understanding. I had mistakenly assumed that he, like others, would be solely interested in tangible outcomes. But Dr. Lukowicz desired more; he sought the core scientific and philosophical underpinnings of AGI.

Dr. Lukowicz's conservative stand against the AGI claim due to a lack of definition of intelligence is well understood. Had I known his interest in deep science, I would have presented the

Challenging Status Quo

Theory of General Intelligence. I'm confident that he might have at least been open to calling it AGI.

The contrast between Etzioni and Dr. Lukowicz is stark. The vast majority in the AGI community mirror Etzioni's engineering view — they want to know the "what." Individuals like Dr. Lukowicz are rare — those who want to understand the "why." As we'll discuss in the next section, there is a major difference between science and engineering.

This episode underscores a broader point: individuals like Bostrom, Tegmark, and Etzioni, while holding prominent status in the AGI and AI world, often lack a true scientific view of the challenges of AGI. Just because someone is "famous" doesn't mean their views are infallible. Science doesn't care about the opinion of the famous and powerful; it cares about evidence, repeatability, and truth.

∞

A pervasive misconception portrays AGI merely as a supercharged version of our current AI systems. This viewpoint misguidedly infers that exponentially augmenting the capacity of a Narrow AI — akin to an advanced chess engine — would eventually metamorphose it into an AGI. However, this assumption is as flawed as believing that relentlessly speeding up a car will eventually morph it into an airplane. While they both function as modes of transportation, cars and airplanes operate based on completely disparate principles. Analogously, AGI and Narrow AI are fundamentally different.

This misconception permeates both the engineering community and the public discourse, serving as the bedrock for the 'emergence theory' — a belief that general or superintelligence will spontaneously arise from narrow AI. Yet,

Science of Intelligence

to my mind, there's a touch of magic to this line of thinking, a form of intellectual leap that lacks rigorous support.

The development of AGI doesn't hinge on the spontaneous emergence from something else but rests on intentional and methodical construction. If we desire AGI, it won't simply materialize — we must purposefully build it from the ground up.

Also, many people believe that AGI is simply an accumulation of different AIs, each with its own specialized function. They imagine AGI as a multi-purpose tool kit, a "Swiss Army knife" of AIs if you will. This perspective falls short because it fails to acknowledge the general aspect of AGI - the ability to learn, understand, and apply knowledge across a vast range of tasks, not just those it was pre-trained on.

∞

AGI's assumed omnipotence is another fallacy. The widespread belief that AGI will possess near-infinite knowledge and capability feeds the fear of a dystopian future where humans are at the mercy of machines. But this assumption misunderstands the nature of intelligence itself. Intelligence doesn't mean knowing and doing everything—it means effectively using available knowledge and abilities to navigate the environment and achieve goals.

Dreyfus, and many others in the AGI research community, might find it hard to envision a machine undergoing a form of 'childhood' or assimilating 'culture'. But what if there was a structured way for AGI to progressively build its understanding of the world, much like a child does? A way for it to form connections, learn from experiences, and even internalize cultural and ethical norms? This isn't just a hypothetical

Challenging Status Quo

scenario. In the upcoming chapters, I'll introduce the concept of the 'Causality Hierarchy', a foundational framework in our Theory of General Intelligence. It's a model that allows AGI to navigate its environment, starting from scratch, and gradually building a rich tapestry of understanding, layer by layer. It's through this hierarchy that AGI can, in a sense, experience its own form of 'childhood' and 'cultural immersion'. But more on that later.

The common thread weaving through these misconceptions about AGI is a lack of a fundamental understanding of the nature of intelligence. Each of these views, whether it's the idea of AGI as a collection of specialized AIs, the assumption of immediate self-improvement, or the belief in its omnipotence, sidesteps the core question: What is intelligence, really?

Interestingly, despite their deviation from the fundamental essence of intelligence, these perspectives have gained significant traction within the AI research community and the public consciousness. This raises an important question: where do these views come from if they are not rooted in fundamental science?

While there's no definitive answer to this, one could speculate that these views may arise from the extrapolation of trends seen in narrow AI, the influence of science fiction, and perhaps a sense of awe or fear surrounding the unknown capabilities of future technologies. However, as we strive towards the development of AGI, it's crucial to return to the bedrock of understanding intelligence itself, divorcing it from popularized misconceptions and forging ahead with insights drawn from a robust, scientific understanding.

Science of Intelligence

Science vs. Engineering

Building on our exploration of the misconceptions surrounding AGI, it's crucial to address another often overlooked aspect: the distinction between science and engineering in the realm of AGI. Oren Etzioni's pointed question, 'What can it do?', which we touched upon earlier, is emblematic of the prevailing engineering-first mindset. While it's a valid query from an application standpoint, it inadvertently sidesteps the foundational question: 'What is intelligence?'.

Sam Altman: "I think - I am very afraid of the fast take-offs. I think in the longer timelines it is harder to have slow take-offs, and there is a bunch of other problems too. But that is what we're trying to do. Do you think GPT-4 is AGI?"

That was Sam Altman's response to Lex Fridman on a question that had nothing to do with whether GPT-4 was AGI or not. The original question Lex Fridman asked had to do with fast take-offs. When we talk about AGI, an intriguing concept often discussed is "fast take-off."

Picture this: You're observing a rocket launch. Initially, the rocket starts off slow, but as it consumes fuel and sheds weight, it accelerates rapidly, soaring high into the sky faster than you can comprehend. That's the essence of a "fast take-off" in the context of AGI.

Now, instead of a rocket, imagine an artificial intelligence system. Today, it's learning basic tasks, understanding the nuances of human language, or solving simple puzzles. However, once this system reaches a certain level of intelligence, it begins to improve itself at an exponential rate. It starts to learn more rapidly, understand more deeply, and solve more complex problems. This is what we call a "fast take-off." It's the moment when an AGI begins to advance its own

Challenging Status Quo

capabilities at such a speed that it far outpaces human comprehension or control.

What puzzled me with Altman's response was the last piece when he asked Lex if he thought GPT-4 was AGI. What did that have to do with fast take-off when it is a well-established fact, even at OpenAI, that GPT-4 is not an AGI?

After Lex gives an answer, which I found puzzling, Altman continues:

"I think GPT-4, although quite impressive, is definitely not an AGI. But isn't it remarkable that we're having this debate?"

"What's your intuition why it is not", Lex asked.

"I think we're into the phase where specific definitions of AGI really matter. Or we just say we know when we see it and we don't bother with the definition.", Altman answers.

I listened to this part of the podcast a few times trying to read between the lines. Is he subtly implying that GPT-4's failure to achieve AGI status is essentially due to our lack of a concrete definition for AGI? Could he be subtly steering the narrative to sculpt the definition of AGI around the principles of large language models (LLMs) — the foundational technology underpinning GPT-4?

In psychology and communication, there is a concept called framing. This is when a subject, in this case, Sam Altman, attempts to plant an idea even though, like Altman, officially denies it. In this case, Altman has publicly stated many times that GPT-4 is not an AGI, but yet in interviews, he is planting this idea that maybe GPT-4 is not, but let's define AGI around GPT-4 – clearly making him the inventor, or the boss of the team inventing history's most impactful technology.

Science of Intelligence

But there is, what I call, scientific blasphemy associated with Altman's comments. First of all, it is not AGI that should be defined. AGI is a machine implementation of intelligence. What needs to be defined is intelligence. This brings us to the blasphemous part of his comment. Science and engineering often overlap in everyday life, but that doesn't mean they are the same thing.

After dissecting the misconceptions surrounding AGI, it's essential to delve into a pivotal distinction often overlooked: the difference between science and engineering in AGI's realm. Oren Etzioni's pointed question, 'What can it do?' is emblematic of the prevailing engineering-first mindset. While it's a valid query from an application standpoint, it inadvertently sidesteps the foundational question: 'What is intelligence?'.

As we navigate the landscape of AGI, it's tempting, as Etzioni's comment suggests, to prioritize the tangible, the engineered outcomes. But this approach, akin to putting the engineering cart before the scientific horse, might be premature. Before we ask what AGI can do, we must first understand what intelligence truly is.

Science is the field of understanding nature, from the tiniest of particles to the cosmos and everything in between, with the exception of artificial objects. It is about understanding how the universe works, the laws that govern it, and its limitations. Engineering is about using the knowledge from science to manipulate, including building stuff, within the laws of nature as we've discovered it.

Intelligence, as a phenomenon, is an integral part of the natural world – whether we perceive it as an emergent property or, as I argue in this book, an element woven into the very fabric of

Challenging Status Quo

SpaceTime. Engineering, invariably, follows science; it stands on the bedrock of our understanding of the natural world. Proposing to define a scientific field around an engineered product is akin to uttering profanities in a house of worship. Esteemed historical figures in science, like Einstein, Bohr, and Planck, would have likely been aghast at such a suggestion.

A similar approach is observable in DeepMind's strategy, another prominent AI research company. Despite its CEO, Demis Hassabis, holding a PhD in neuroscience, much of DeepMind's work appears to be focused on developing advanced AI models, rather than understanding the phenomenon of intelligence itself. While neuroscience undoubtedly provides valuable insights into human cognition, it primarily deals with the mechanics of the brain, not the broader science of intelligence.

By framing the quest as "Solving Intelligence", DeepMind seems to be putting the engineering cart before the scientific horse. This might make for impressive technological achievements in the short term, but could ultimately hinder our progress towards AGI.

Hassabis's recent announcement confirming DeepMind's intention to create an algorithm far surpassing GPT-4's capabilities underlines this focus on engineering (Knight, 2023). While this promises to be an impressive product, it's imperative to remember that this endeavor remains firmly within the realm of engineering. Even with substantial advancements, it doesn't significantly shift our position towards achieving AGI.

In an entertaining exercise, I asked GPT-4 to assume the persona of Albert Einstein, armed with the wealth of information we have about him. I then requested it to write a

Science of Intelligence

letter to Mr. Altman expressing its views, having heard his comment. The result, unedited, is attached to Appendix B. I encourage you to take a pause and enjoy this letter in the appendix.

I recently had the pleasure of reading Walter Isaacson's biography of Einstein, and I must say that I was taken aback by how closely the letter composed by GPT-4 mirrored the tenor and content of Einstein's historical letters as presented in the biography. This in itself is a tribute to the astonishing achievements of Altman and his team at OpenAI. However, it's worth emphasizing that the brilliance of GPT-4's mimicry does not absolve the problematic nature of Altman's comments.

Reflecting back on Oren Etzioni's question, "What can it do?", it becomes clear that this engineering-centric perspective, while valuable in its own right, can sometimes overshadow the foundational scientific inquiries. While it's essential to understand what a system can achieve (engineering), it's equally, if not more, crucial to understand the underlying principles that enable such achievements (science). In our quest for AGI, we must ensure that we're not merely building impressive tools without a deep understanding of the very phenomenon we aim to replicate: intelligence.

Faking Intelligence

Some proponents in the field argue that it's enough to engineer an AI system that can convincingly "fake" intelligence. To a degree, this argument isn't without merit. For example, OpenAI's GPT-4, the latest in a line of increasingly sophisticated language models, has demonstrated an extraordinary capacity to generate human-like text. In fact, GPT-4 even assisted in the development of this book, providing valuable insights and generating articulate narratives.

Challenging Status Quo

Demis Hassabis have promised another chatbot that would far exceed the capabilities of GPT-4. However, is GPT-4 or Bard, Google's version of a chatbot, intelligent?

Philosopher John Searle's Chinese Room thought experiment is quite illustrative here (Searle, 1980). In the experiment, Searle asks us to imagine a person in a room who follows English instructions for manipulating Chinese symbols. To those outside the room, it appears that understanding of Chinese is occurring within the room. However, the person inside the room doesn't understand Chinese at all; they're just following instructions. In essence, Searle argues that a computer could appear to understand language, or simulate any other aspect of human cognition, without actually understanding.

GPT-4 operates similarly. It generates text that appears intelligent by predicting the next word in a sequence based on patterns it has learned from a vast dataset of human-written text. The resulting output can be impressively coherent and contextually appropriate, which certainly gives the impression of intelligence.

But here's the crux: GPT-4 doesn't truly understand the content it generates or processes. It manipulates symbols (words) based on learned statistical patterns without any semantic understanding of the words it uses. It doesn't retain memory of past interactions beyond the immediate conversation and doesn't form beliefs or make conscious decisions. Like the person in the Chinese Room, GPT-4 is a highly advanced symbol-manipulating machine, not a thinking entity.

While GPT-4 and its ilk can convincingly "fake" aspects of intelligence, they fall short of genuine comprehension, consciousness, and flexible, adaptive learning — the essential

Science of Intelligence

hallmarks of real intelligence or AGI. This doesn't discount the impressive technological achievement that GPT-4 represents or its practical utility. However, it is crucial to differentiate this form of artificial "intelligence" from the more profound, comprehensive concept of genuine intelligence or AGI.

Consequently, even as we marvel at the remarkable capabilities of GPT-4 and similar AI systems, we must not let their simulated "intelligence" mislead us. Real AGI — a system that truly understands, learns, adapts, and perhaps even experiences — will not emerge merely from refining these pattern-matching techniques. Genuine intelligence is not a veneer of sophisticated responses; it's a deep, intricate tapestry woven from threads of understanding, learning, adaptation, and perhaps even consciousness.

A machine that can truly learn to perform tasks that we, as humans, have yet to master requires a depth of understanding and capability that goes beyond mere mimicry. And this, I argue, requires us to transcend our current engineering-focused approach and truly grapple with the science of intelligence.

Chapter 3

Journey to Understanding Intelligence

"In questions of science, the authority of a thousand is not worth the humble reasoning of a single individual."

Galileo Galilei

Science of Intelligence

In my earliest explorations of artificial intelligence, one certainty steadfastly anchored my beliefs: speaking to a VCR or any rudimentary voice interface could not, by any stretch, be considered authentic intelligence. Fast forward to the modern era, where we casually interact with Siri, Alexa, and Bixby, collectively branding them as 'artificial intelligence.' Yet, even now, my conviction remains unaltered - these are not manifestations of genuine intelligence.

Siri was launched with much fanfare by Apple in October 2011 as part of their iPhone 4S announcement. As one of the earliest mainstream voice assistants, Siri paved the way for a new form of interaction with our devices, sparking both fascination and skepticism in equal measure. Amazon Alexa, introduced in November 2014, expanded the concept further, moving beyond smartphones to becoming a standalone device designed for home use, known as the Amazon Echo. The device made waves, promising to be a new type of smart home hub. Not far behind was Samsung's Bixby, unveiled in March 2017, as part of their Galaxy S8 smartphone. Although a latecomer, Bixby offered deep integration with Samsung's devices, aiming to provide a more seamless user experience.

While the marketing of these virtual assistants was couched in terms of "artificial intelligence," the reality, as always, proved more nuanced. These technologies employed advanced natural language processing, machine learning, and vast amounts of data to understand and respond to user requests. However, they fell short of exhibiting autonomous intelligence, remaining fundamentally bound by their programmed algorithms and databases.

But for many users, these distinctions mattered little. The ability to converse, however artificially, with our devices was

Journey to Understanding Intelligence

seen as revolutionary. It made me wonder: if we're so captivated by these early iterations of "intelligence," how would we react to truly advanced, almost unimaginable technologies? This line of thought naturally led me to the fringes of our technological imaginations, where the lines between science, fiction, and folklore blur.

It was during this period of contemplation that I stumbled upon the world of UFO conspiracy theories. As a natural skeptic, I viewed these theories as intellectual exercises, rather than concrete truths. However, it was during a television documentary on this very subject that a particular concept caught my attention. The documentary presented an alleged extraterrestrial device, a hypothetical instrument purportedly capable of predicting the future. As fantastical as it seemed, this idea sparked a wildfire in my mind: If such a device could predict the future, surely it held the potential to change it? But how?

My initial reaction was skepticism. Could such a device even be theoretically possible, outside the realm of UFO documentaries and science fiction? It didn't take long before I realized that according to Einstein's relativity theory, combined with Einstein's view of a deterministic block universe, such a device wasn't entirely beyond the realm of possibility.

This foray into the theoretical implications of advanced technology steered me towards studying physics. While this diverged from the prevailing views on intelligence both then and now, I deemed it an essential detour. Yet, this perspective introduced its own set of foundational questions.

My thinking began to coalesce around the concept of free will. The device unnerved me, prompting a cascade of profound questions. If this extraterrestrial technology could indeed forecast the future, what implications did this bear for free will? If we are not the sovereigns of our destiny, does intelligence, as we perceive it, truly exist?

Eager to bridge the divide between free will, intelligence, and the deterministic outcomes observed in physics, I delved deep into existing theories. What I hadn't anticipated, however, was that this pursuit would lead me to forge my own interpretation within quantum mechanics. But before diving into that, let's first unpack the prevailing wisdom in the realm of physics.

Intelligence in a Deterministic SpaceTime

If we were to draw back the curtain of reality and peek at the inner workings of the universe, what would we find? A universe of deterministic laws or one of free will and chance? This fundamental question has intrigued and divided thinkers for centuries. On the one hand, the laws of physics as we currently understand them paint a picture of a deterministic universe, one in which the state of everything at one moment entirely determines its state at the next. This view is known as determinism, and it essentially posits that the universe operates like a vast, intricate clockwork mechanism, with each part ticking along in perfect, predictable harmony with the others.

Brian Greene, a renowned physicist and author, adopts this deterministic perspective (Greene, 2004). He postulates that our universe operates on the principles of quantum mechanics, which, despite their probabilistic nature at the microscopic level, yield a largely deterministic universe at the macroscopic level. However, he also acknowledges that we have the feeling of free will—of making choices. This sense of agency, according to Greene, emerges from our limited perspective: we can't perceive or compute all the variables that influence our actions, leading us to perceive our actions as freely chosen.

The deterministic perspective can be unsettling. It reduces life and all its complexity, beauty, and mystery to a kind of cosmic movie, one in which the script has been written, the scenes set, and the characters' actions predetermined. No matter how diligently we study, how creatively we think, how passionately we love, or how boldly we dream, it's all part of the script—a script we didn't write and cannot alter. As Greene suggests, we may believe we're free because we don't see the prison bars.

Yet, I found this line of argument unsatisfactory. Extending Greene's prison bar analogy, he is essentially arguing that if a prisoner is oblivious to their confinement and perceives themselves as free, then they might as well be free. This mirrors Einstein's assertion that free will is an illusion, implying that the perception of freedom (for the prisoner) is likewise an illusion. I found this philosophically untenable.

Despite this, Greene and other scientists often emphasize that this deterministic perspective doesn't necessarily diminish the reality of our experiences or the validity of our human-level concepts. Even if our sense of free will is not grounded in fundamental physical laws, it's still a crucial aspect of our human experience and our understanding of ourselves.

Philosophically, many have grappled with the implications of determinism. Renowned physicist Albert Einstein, a proponent of determinism, believed in the causality principle of physics. However, his thoughts on the implications of determinism on personal accountability and moral responsibility are subject to interpretation. Nonetheless, the question remains: If our actions are predetermined, can we be held morally or ethically responsible for them? This quandary, while reasonable in a

Science of Intelligence

deterministic universe, rouses profound debates on morality, ethics, and justice.

The debate between determinism and free will is not just philosophical but also profoundly influences how we understand intelligence. If we lean towards a deterministic universe, then intelligence seems like a moot point. There's no purpose to life or the universe if we're just a movie playing. But if we see some room for free will, then the concept of intelligence gains significance—it becomes a vital instrument in navigating the unfolding movie of life, a tool that helps us make decisions, solve problems, and pursue our goals.

The good news is that I saw a hole in the views of people who subscribed to the deterministic view. They argue for determinism because every experiment in physics, once we understand the laws governing the experience, yield the predicted result every single time.

Experiments are not testing consciousness decision making, or free will. The question I started thinking about was who initiated the experiment? Are we predestined to make that experiment and see that expected result. According to Einstein and Greene, we are. But I didn't buy that answer.

The only thing I knew was that I had an urge to believe that free will exist, otherwise it appeared to me that my research had little value.

Intelligence and Quantum Mechanics
While I entertained suspicions of a possible loophole, the deterministic universe, staunchly championed by Einstein, appeared to leave a scant margin for the operation of free will.

Journey to Understanding Intelligence

However, the enigmatic world of quantum mechanics might hold a key to unlocking a different perspective.

From our earliest education, we comprehend that all matter is composed of atoms and molecules, which in turn consist of particles such as electrons, protons, and neutrons. Quantum mechanics—the physics that elucidates the behaviors of these particles—poses bewildering challenges to our common intuition. One of the perplexing questions that arises is, "Where is the electron?" Unveiling the answer takes us into the heart of quantum mechanics and introduces us to the wave function.

The wave function is a mathematical concept, but let's break it down into simpler terms. Imagine you're looking for your pet in your house. You can't see it, but you know the places it likes—the cozy corner in the living room, under your bed, near the window sill. You can make educated guesses on where your pet might be but can't be certain until you actually look.

Similarly, in the quantum world, the wave function helps us make educated guesses about where an electron might be found. Instead of a specific location, it provides a range of probabilities of finding the electron in different positions. An electron doesn't have a defined place until we measure or observe it. Until that moment, it exists in a state of possibility—being everywhere and nowhere at once, within a defined space.

To illustrate further, an electron in an atom doesn't exist as a distinct point in space but as a 'cloud' of probabilities. This 'cloud' isn't a physical entity but a mathematical expression of all the places the electron could potentially be, each with a certain probability. The precise location of the electron remains undefined until we measure it, compelling us to

Science of Intelligence

perceive its existence through a lens of probabilities, rather than certainties.

This inherent probabilistic character of particles injects a degree of uncertainty into the realm of quantum mechanics. To grapple with these peculiar results, scientists have devised multiple interpretations. Although these interpretations do not qualify as verifiable theories, they offer narrative-like frameworks that enable us to comprehend quantum phenomena.

Copenhagen Interpretation

The Copenhagen interpretation emerged in the early 20th century, primarily developed by Niels Bohr and Werner Heisenberg, two of the era's most notable physicists, in the city of Copenhagen. It marked one of the pivotal points in the evolution of quantum theory, introducing a conceptual framework that was profoundly different from classical physics.

Bohr and Heisenberg proposed a view of quantum mechanics that was steeped in probabilistic outcomes and uncertainties. At the core of this interpretation lies the "wave function," a mathematical entity that encapsulates the probabilities of finding a particle, like an electron, in various positions. However, this interpretation also brought to light a vexing question: "Where is the electron?" In the context of the Copenhagen interpretation, this question didn't have a definitive answer until a measurement was made.

The Copenhagen interpretation posits that an electron pervades all space simultaneously until measured, existing in a superposition of all possible states. The act of measurement 'forces' the electron into one of these states; it's here that the wave function "collapses" into a single point, indicating the electron's position. This process infuses an element of randomness into the fabric of reality — a randomness that is foundational and irreducible.

This idea was a departure from the deterministic view of the universe, where every event is considered to be definitively predictable. Einstein famously expressed his discomfort with this notion, encapsulated in his phrase, "God does not play dice." Yet, the Copenhagen interpretation has persisted as a cornerstone in the teaching and understanding of quantum mechanics.

Educational institutions, from high schools to universities, predominantly favor teaching the Copenhagen interpretation. This interpretation, with its inherent randomness, poses an existential challenge to the notion of free will. It raises the complex question of how we can exercise control over our actions if the fundamental particles and forces governing the universe are inherently unpredictable.

Yet, it's essential to note the ongoing debates. Numerous scientists and philosophers harbor discomfort towards the Copenhagen interpretation. The quantum world is still a landscape of mystery and exploration, and questions like "Where is the electron?" continue to provoke profound inquiries into the nature of reality, observation, and existence.

The inherent randomness introduced by quantum mechanics inevitably leads us to question how our macroscopic world maintains its stability and predictability. How do buildings stand, planets orbit, and life exist in a universe built from quantum blocks dancing to the tunes of probability?

Science of Intelligence

Many-World Interpretation

Hugh Everett III, an American physicist, ventured into the world of quantum mechanics when the field was still in its relative infancy. Graduating from Princeton University in 1953, Everett conceived of the groundbreaking Many-Worlds Interpretation (MWI) while he was still a doctoral student working under the renowned theoretical physicist John Archibald Wheeler.

In his 1957 Ph.D. thesis, "The Theory of the Universal Wave Function," (Everett, 1957) proposed a novel and audacious departure from the Copenhagen Interpretation, the then-dominant understanding of quantum mechanics. Instead of subscribing to the concept of the wave function collapsing into a single outcome during measurement, Everett suggested a radical alternative: each possible outcome of a quantum event happens but in a different 'world' or universe.

This brings us to the haunting question, "Where is the electron?" Under the MWI, every potential location of the electron, according to its wave function, is realized in its own separate branch of the universe. There is no collapse of the wave function, no single outcome; instead, every possible outcome occurs, each in its own "world." It suggests a multiverse where every quantum possibility is realized. In one universe, the electron is found here; in another, it is there, and in yet another, it's somewhere else entirely.

In simpler terms, this suggests an almost infinite number of universes existing in parallel, each reflecting a different outcome of every decision you've ever made. So, for instance, in one universe, you might be an award-winning actor, while in another, you're a successful scientist. There could even be a universe where you are the President of the United States, and yet another where you have chosen a path of crime and are sitting on death row. This staggering array of possibilities is as fascinating as it is mind-boggling.

The MWI represented a profound shift in understanding quantum mechanics. It transformed the abstract and often counterintuitive theories into a picture of reality that seemed more at home in the pages of a science fiction novel than in a physics laboratory. Nonetheless, it opened the door for an entirely new way of understanding our cosmos and the fundamental nature of reality. This proposition was met with skepticism and even mockery from the mainstream physics community. Some deemed it an extravagant and unnecessary complication to the already complex field of quantum mechanics. Niels Bohr, one of the founders of the Copenhagen interpretation, rejected Everett's ideas outright, leading to significant disillusionment for Everett.

However, as the years passed and the quantum conundrum deepened, Everett's MWI began to attract a considerable following. Its elegance and consistency offered a promising alternative to the conceptual difficulties and ambiguities associated with the Copenhagen Interpretation. Gradually, more physicists, dissatisfied with the metaphysical aspects of the Copenhagen interpretation, began to take Everett's ideas seriously.

By the late 20th century, the MWI started to gain acceptance and even became the subject of serious scientific discussion. Notable contemporary physicists, such as Stephen Hawking and Sean Carroll, have since publicly voiced their support for the MWI. Today, despite its unconventional implications, the Many-Worlds Interpretation is one of the most popular interpretations among physicists and is regarded as a viable and cogent explanation of quantum phenomena.

Science of Intelligence

Everett's story serves as an exemplary reminder that innovative ideas often face initial resistance but can eventually revolutionize our understanding of the universe when given the room to grow and the courage to be examined seriously.

Delving deeper into the implications of the MWI, it postulates that all potential outcomes of quantum measurements manifest in distinct universes. Each quantum event instigates a "split" in the universe, culminating in multiple realities, each corresponding to a different outcome. For instance, upon arriving at a junction, your choice to veer left or right engenders two universes—one where you swerved left, and another where you turned right.

∞

It's a crazy concept, isn't it? Despite its audacity, this perspective of quantum mechanics attracts the allegiance of several eminent physicists worldwide.

While this interpretation has inspired myriad science fiction narratives, including Star Trek's alternate universe episodes, it also provokes philosophical inquiries about free will.

Prominent theoretical physicist Sean Carroll, an advocate of MWI, regards free will as an emergent phenomenon that harmonizes with this interpretation. He proposes that while every conceivable outcome transpires in some branch of the universe, we execute choices within our branch (Carroll, 2019). Our choices, governed by our internal states and the laws of physics, impart the sensation of free will. The existence of other branches, where we opted for different decisions, does not detract from the authenticity of our experiences or the relevance of our decisions in this universe.

Journey to Understanding Intelligence

However, if every decision we might make materializes somewhere in the multiverse, it could insinuate that our liberty to choose is supplanted by the inevitability of all choices coming to pass. If every decision manifests in some universe, it appears to subvert our capacity to 'choose not to do something'—an integral aspect of free will.

Both the Copenhagen interpretation and the MWI present compelling possibilities and pose challenges to our comprehension of the universe and our position within it. Quantum mechanics, in all its myriad interpretations, does not proffer an unequivocal resolution to the enigma of free will or the function of intelligence. It leaves us in contemplation about the inherent randomness or the multiplicity of universes and what implications they bear on our perception of free will.

Quantum Bayesianism (QBism) and Free Will
There's one interpretation of quantum mechanics that seems to directly grapple with the concept of free will—Quantum Bayesianism, or QBism. Emerging from the work of physicist Christopher Fuchs and his colleagues in the late 20th and early 21st centuries (Fuchs, 2014), QBism presents an alternative view of quantum mechanics, positioning it from a subjective standpoint.

In QBism, the wave function is treated as a predictive tool for estimating the probabilities of different outcomes, rather than a depiction of objective reality. The wave function embodies an individual's subjective belief about the possible results of an experiment, suggesting that quantum mechanics describes not the world as it is, but how we interact with it.

This subjectivity directly addresses the question, "Where is the electron?" In QBism, the answer is inherently tied to the observer's beliefs. If you believe the electron is more likely to

Science of Intelligence

be found in a particular location, then the wave function is tailored to reflect this belief. It means that each observer can have their own wave function for the same system, shaped by their prior knowledge and beliefs. The act of measurement then serves as an update to these beliefs. The electron's position isn't uncovered but instead becomes part of the observer's updated set of beliefs.

When it comes to free will, QBism offers a compelling standpoint. It underscores the experimenter's autonomy to select which measurements to make. This freedom, according to QBism, isn't just an illusion; it's a fundamental feature of the universe. Since the wave function mirrors the experimenter's expectations and beliefs, the concept of choice becomes ingrained in quantum mechanics.

Christopher Fuchs frequently emphasizes this perspective. He contends that free will doesn't clash with the laws of physics—instead, it's essential for deciphering the quantum world. QBism acknowledges our incapacity to predict certain occurrences, attributing this not to our lack of understanding, but to the inherent nature of the universe.

However, Sean Carroll, a physicist well-versed in the nuances of quantum mechanics, harbors reservations about QBism's inherent subjectivity. At the heart of his criticism is a foundational concern: quantum mechanics, though undeniably strange and counterintuitive, should still aspire to describe some objective facet of reality. By anchoring the wave function—a cornerstone of quantum theory—in the beliefs and expectations of the observer, QBism seems to drift away from this objective.

Carroll posits that a successful interpretation of quantum mechanics should provide a consistent narrative, wherein the universe follows a set of discernible, objective rules. This doesn't mean that everything must be deterministic or predictable, but rather that the fundamental processes governing the universe should not be contingent on individual beliefs. If QBism's subjectivity is taken to its logical conclusion, it might imply that there's no underlying quantum reality independent of our interactions or beliefs—a radical departure from the scientific approach of uncovering objective truths.

Yet, QBism is not without its critics who challenge its subjective nature. They contend that QBism undermines the basic premise that the universe exists independently of our beliefs or knowledge. They also point out that QBism's reliance on the notion of free will contradicts the deterministic framework of classical physics.

Despite these ongoing debates, QBism underlines the unresolved tension between determinism and free will in quantum mechanics. It calls into question our comprehension of reality and nudges us to reassess our assumptions about the nature of the universe and our role within it. Though I commend the attempt to integrate free will, I find myself aligning with Carroll's perspective. The notion of introducing subjectivity into a universe governed by natural laws does seem to raise valid concerns.

Flexible Block Universe and Quantum Mechanics

Nothing I came across satisfied my need to unify free will and intelligence with the innerworkings of the universe. If the universe were deterministic, as both Einstein's relativity and prevailing quantum mechanics interpretations seemed to suggest, then the notion of intelligence might be a moot point. Why study the nature of intelligence if our decisions, our thoughts, even our consciousness are preordained? This

Science of Intelligence

quandary sent me on a quest into the depths of quantum theory, a journey that culminated in the creation of a new interpretation, one I'm sharing publicly for the first time. I acknowledge it might be mind-bending, but I invite you to keep in mind that the merit of an interpretation lies in its alignment with empirical data, a test that my proposition, the Flexible Block Universe, does not fail.

So, let's begin. Before we can delve into this new interpretation, we need to grasp the concept of the block universe, or Eternalism. Eternalism is a philosophical approach to the nature of time, which posits that all points in time are equally "real," as opposed to the presentist view that only the present moment is real. Einstein's theory of relativity supports this view, postulating that time is a dimension like space, and all events in time exist in a timeless, unchanging four-dimensional "block."

However, this mainstream interpretation leans heavily on an assumption that is not empirically rooted—that both the past and future are fixed and deterministic. While it may seem natural to consider our past as a fixed record, this is more of a reflection of how our minds process, store, and recall information than a fundamental feature of the universe.

Similarly, the idea of a fixed future may stem from our limited perspective as creatures bound by time. In our everyday experience, the outcomes of our actions may seem predetermined because we can only act based on the information available to us at the time. However, this does not necessarily indicate that the future is already set in stone.

The truth is, there is no theoretical or mathematical reason to treat the past and future as fixed entities. Even the equations we've accepted in physics do not distinguish between the past

Journey to Understanding Intelligence

and future. Therefore, this idea of a rigid timeline is more of an interpretation than a factual assertion about the nature of the universe.

While the block universe interpretation holds its own scientific merit, it seemed, to my mind, overly rigid. This rigidity seemed to deny the universe its innate dynamism and playfulness, especially in the context of free will. Consequently, I found myself asking: Could it be that the block universe isn't a static, immovable block, but rather a flexible, mutable entity?

These contemplations gave rise to a concept that I eventually named the Flexible Block Universe (FBU). In this proposed interpretation, the universe is still a block in the sense that all moments in time are real. However, in contrast to the conventional block universe, these points in time are not set in stone. They possess a degree of malleability, subject to change and evolution. Thus, the universe under this perspective resembles a vast ocean: fluid, dynamic, and perpetually in motion.

Let's visualize this concept. Imagine if you could momentarily step outside our block universe. For the sake of this thought experiment, let's overlook the reality that no current physics theories would permit this. From this external viewpoint, you'd see the entire 4D universe unfolding. At one end, the Big Bang ignites, paving the way for our solar system and Earth. As your gaze traverses this 4D expanse, you'd witness the reign of dinosaurs, the emergence of humanity, and finally, at the distant end, the culmination of time. All these moments coexist concurrently.

To simplify, let's strip the universe down to a single particle. This particle has existed, exists now, and will continue to exist—from the Big Bang to the end of time. What you would see is not a minuscule point-like particle. Instead, your eyes would trace a string extending from the Big Bang to the brink of eternity.

To bring this idea into sharper focus, I frequently present my students with a thought experiment. I hand them a long rope, with each end held by a student. One end signifies the Big Bang, the other the end of time. This rope emulates the four-dimensional particle stretched across time. Marking a spot on the rope symbolizes a slice of 'now'—our three-dimensional perception of the particle.

As the students oscillate the ends of the rope, waves ripple along its length. Trying to pinpoint where the 'present particle' is now becomes a complex question. The 'present particle,' much like a wave's superposition, may appear to be in numerous places at once, depending on the speed and pattern of the rope's motion. Should we seek to measure the particle's exact location, we'd need to hold the rope, thereby dampening the waves at that spot. Our measurement, interestingly, could affect other parts of the rope, influencing both past and future.

In the FBU context, the "Where is the electron?" question gains a new dimension. The electron's position is real for every now-slice, but it isn't static. It keeps changing, influenced by the wave ripples of spacetime created by its movement in both the past and the future. While it is not measured, the electron is free to follow these wave ripples. However, once we measure it, its position becomes fixed, and this act of fixation isn't isolated—it affects the electron's position in the immediate past and future due to the interconnected, dynamic nature of spacetime in the FBU.

The FBU interpretation suggests that the universe is not a static entity, as the traditional block universe postulates. It's also not purely random, as suggested by the Copenhagen interpretation, nor does it split into separate universes with every quantum event, as the Many-Worlds interpretation claims.

Instead, the FBU is flexible and dynamic. It behaves like a block of jello. When you poke it (i.e., when a measurement or an event occurs), you create a ripple that travels both forwards and backwards in time. Therefore, in this interpretation, what you do now changes your future and also your past. This understanding adheres to all known empirical data from quantum mechanics and opens the door to free will.

To some, the idea that your actions now affect what you did an hour ago might sound counterintuitive. However, consider this: if you throw a rock onto a body of water, ripples spread in every spatial direction—left/right, forward/backward, and even upward (seen as the amplitude of the ripple). We readily observe these spatial perturbations. In essence, we observe 3.5 dimensions—the three spatial dimensions and the forward progression of time. If, as suggested by general relativity, time is just another dimension, it's logical to propose that it has two directions: forward and backward. Just because our typical experience limits us to observing only the "forward" half of the temporal dimension doesn't mean the other half doesn't exist or behave similarly. Like the stone causing ripples to propagate outward in water, events might ripple through time, influencing both the future and the past.

The FBU view allows for an interesting perspective on consciousness. Imagine that a human being, comprised of trillions of particles, could be represented as a bundle of these 'strings.' Within this framework, countless versions of 'you' would stretch into both the past and the future. Each temporal

Science of Intelligence

slice of you might experience reality slightly differently based on their unique point in time. Yet, consider the possibility that the cumulative experiences of these slices—the entirety of these moments—is what actually constitutes the "real you". Could this collection of experiences form the foundation of the consciousness that represents you?

Each "slice" of you would, in essence, be its own consciousness, or what we might call an "agent", capable of decision-making. The free will exercised by one of these slices would be inherently limited due to its temporal constraints. However, the free will of the entire 4D-you—the aggregation of all these slices—could possess far broader autonomy.

This perspective unfurls intriguing possibilities concerning free will. It suggests that each individual slice might be acting in service to a grander objective, one that no single slice could fully fathom on its own.

Introducing consciousness into the block universe might sound far-fetched, but it's a concept that some mathematical models can accommodate. Notably, Judea Pearl introduced the do-operator, a mathematical operator that embodies the idea of imposing a change on reality (Pearl, 2018). In a deterministic world, the do-operator would be redundant, but in the FBU, it symbolizes free will.

The FBU introduces another tantalizing speculation: if the past is in flux, why don't we observe our memories changing? The answer lies in our perception and personal experience of reality. Our memories are not live feeds of past events; instead, they are recordings of our personal causal chain of events.

Even though this chain may have shifted in the 'actual' past, our memories remain untouched. The past we remember is

Journey to Understanding Intelligence

our past – the one we lived through, not the one that currently exists in the spacetime fabric. This discrepancy suggests that our individual perception of reality might be fundamentally different from the true state of the universe according to the FBU. It underscores the uniqueness of our subjective experiences, hinting at a deeper layer of complexity in understanding consciousness and free will.

While these ideas are mind-bending and challenge our conventional understanding of reality, they're essential in our quest to comprehend the nature of intelligence. It's only by pushing the boundaries of our understanding that we can hope to unravel the complex interplay between determinism, randomness, and free will in the universe.

This living, dynamic universe doesn't just involve the particles. It involves you, and every other conscious being that exists. You see, in the FBU, agents play a pivotal role. They can 'poke' the universe, causing ripples that travel in all directions across spacetime. These ripples embody change, evolution, and growth.

Consider this: if we were to remove all conscious beings from the FBU, it would revert to being deterministic, much like Einstein's static Block Universe model. Without agents to make choices, to apply the 'do' operator, the universe would have no reason to change. This highlights the vital role we play in the FBU. Our choices, actions, and existence influence both the future and the past, carving out a unique path in the block universe.

This is a universe where free will is not only possible, but integral to its very structure. And that makes it far more conducive to the concept of intelligence than a deterministic universe. After all, without the freedom to make choices and effect change, intelligence loses its purpose.

While the FBU interpretation might be met with skepticism due to its divergence from conventional physics, it introduces concepts like retrocausality that may initially be challenging to grasp. Yet, it's crucial to underscore that, even with its unconventional premises, the FBU doesn't contravene any known laws of physics or dismiss any empirical data. Instead, it offers a fresh lens through which to interpret this data, paving the way for a universe where intelligence has a meaningful role, something that current mainstream interpretations often struggle to accommodate.

Self-Reflection: Intelligence & The Cosmos
The FBU interpretation might seem unrelated to our exploration of intelligence. However, I see it differently. For me, a universe that allows for the existence of free will and intelligence was a prerequisite for delving into the science of intelligence. Without such a universe, studying intelligence seemed futile - it would merely be a human-made construct devoid of any inherent significance.

In my view, FBU did more than just accommodate intelligence; it harmonized all intelligence across the cosmos. By positing that extraterrestrial intelligence may also exist, it instills a sense of unity, suggesting we are all connected in the grand scheme of the universe's evolution – we are integral parts of the universe itself. This realization imbues us with a sense of purpose and inspires the belief in something larger than ourselves.

Journey to Understanding Intelligence

I hold that the ultimate goal of science should be to answer the question, "Why do we exist?". This doesn't mean every scientific pursuit should be subordinate to this philosophical query, but rather it serves as a broader guiding principle that lends strategy to our scientific 'tactics'. As the ancient Chinese military strategist Sun Tzu stated, "Strategy without tactics is the slowest route to victory. Tactics without strategy is the noise before defeat." If philosophy is the strategy and science is the tactics, this shows that they cannot be decoupled as we explore the existence of reality.

It's conceivable that 'Why do we exist?' is not the ultimate question, but whatever that ultimate question might be, it provides a compelling motivation for our collective endeavor.

In a deterministic universe, as portrayed by Einstein and Greene, where all of our actions and feelings are predetermined, they still argue for the existence of love, beauty, and art.

I must respectfully disagree with them, from my perspective, on this point. If my emotions, the person I'm going to fall in love with, or the art that moves me, are all predestined occurrences, then in my view, they lose a crucial element of their authenticity. They may appear real to us in the moment, and for many, maybe that is enough. But for me, they cannot be truly real without the element of free will, choice, and unpredictability. This is not to say my interpretation is absolute, but it's a perspective I firmly hold.

FBU offers me that solace, the assurance that the universe has a real purpose, that love and art are genuine - not illusions. FBU also suggests that there are grander mysteries awaiting discovery, potentially bringing us closer to understanding life's purpose and the mysteries of the universe itself.

Science of Intelligence

Echoing Einstein's sentiment in his letter to Thornton, I observe that many scientists seem to focus predominantly on empirical evidence and methodologies, often neglecting the philosophy of science. I see this as a missed opportunity. To me, scientists should resemble detectives, not just collecting evidence, but also trying to piece together a larger narrative.

Recognizing that we don't yet have answers to some of the most profound questions in science, I see a sense of purpose as an indispensable element of our collective scientific journey. Certainly, there are concrete, immediate purposes such as 'figuring out the nature of particles to build better transistors', but these, to me, are mechanical purposes, nested within a larger, more meaningful quest.

Science, in my view, should be motivated by more than immediate, practical outcomes. It should be a pursuit towards understanding the deeper realities of our existence, and ultimately, addressing that grand question – why do we exist? Granted, we may not be able to measure or describe this purpose in tangible terms at the moment, but the act of seeking it gives our scientific endeavors a sense of direction and profundity – or, as Einstein described it, makes us the seeker of truth.

This viewpoint naturally extends to the subject of intelligence. FBU was never meant to be a detached philosophical detour from my work on intelligence; rather, it was intended to bring new meaning to it.

In the context of FBU, intelligence holds a prominent position. The universe can be viewed as being 'alive' through the consciousness that exists across the cosmos, facilitating its evolution.

Journey to Understanding Intelligence

This led me to a realization: intelligence, as a concept, is often diffused across various fields of study, from neuroscience to computer science, but rarely studied on its own terms. I firmly believe it should be recognized as its own field — the science of intelligence.

This perspective has not only invigorated my research but also imbued it with a renewed sense of purpose.

Chapter 4

The Theory of General Intelligence

> "The beauty of a living thing is not the atoms that go into it, but the way those atoms are put together."
>
> *Carl Sagan*

The Theory of General Intelligence

Now that I was confident the universe accommodated intelligent entities, it was time to delve deeper into the essence of intelligence. Scientists generally concur on one aspect about intelligence: it fundamentally involves achieving goals. However, just as with intelligence, we lack a universally accepted scientific definition of 'goals'.

What struck me was the seeming paradox of this situation. We link intelligence—an elusive, undefined concept—to the achievement of goals, themselves abstract and varied entities. To me, this revealed an astonishing gap in our understanding. How could the field of artificial intelligence, after decades of exploration, still grapple with such fundamental concepts?

This realization was the impetus for my journey into the science of intelligence. From my perspective, it was a clear indication that we were too entangled in human-centric concepts—staring at a single tree, if you will—to appreciate the entire forest. The field's fixation on human intelligence seemed to have hindered our ability to understand the universal embodiment of intelligence. We needed to disentangle humanity from intelligence to gain a holistic view.

That's where the Theory of General Intelligence comes in. As far as I'm aware, it's the first theory that attempts to define intelligence in a universally applicable way. It shifts our gaze from the individual tree to the grandeur of the forest, viewing intelligence from the perspective of the cosmos rather than humans.

This chapter is dedicated to dissecting and interpreting this theory for the general reader, with the full paper included in Appendix A for those desiring a more in-depth exploration. Our journey begins by unraveling the concept of a 'goal', for if intelligence is fundamentally about achieving goals, it's essential to have a crystal-clear understanding of what these targets might be.

To truly grasp the essence of a 'goal', we must first understand the very fabric of our reality. At the heart of every phenomenon, every object, and every thought lies a fundamental truth: the intricate dance of particles. Before we delve into the abstract realm of goals and intelligence, let's ground ourselves in the tangible, physical world that surrounds us. Let's explore the universe's most basic building blocks and how they shape everything we know and experience.

The Fundamental Reality of Particles

In the moments following the Big Bang, the universe was incredibly hot and dense, filled with a mix of quarks and gluons. These tiny particles, the most basic building blocks of matter, were not yet grouped together as they are today.

But as time moved on and the universe expanded and cooled, these quarks began to group together, forming protons and neutrons. These would become the core of atoms, the basic units of all matter.

Then, about 380,000 years after the Big Bang, the universe had cooled down enough for electrons to join these protons and neutrons, creating the first atoms, mostly hydrogen. This was a major step, setting the stage for the stars, planets, and galaxies we see today.

Throughout all of time, from the moments after the Big Bang to our present day, these basic particles have been a constant

The Theory of General Intelligence

presence. They're in the stars above us, the earth below us, and even inside us.

Building on this idea, it's fascinating to realize that everything we see, feel, and talk about is based on specific groupings of these particles. Every object, every sound, every emotion can be traced back to a unique combination of particles in a particular place and time.

For instance, consider the word "banana." At face value, it's a simple term we use to describe a familiar fruit. But what it truly represents is a shorthand for the specific particle configuration that makes up a banana. It's a simplified way of referring to the myriad of particles arranged in a particular pattern, giving rise to the color, texture, taste, and nutritional value we associate with a banana.

Even the words we use, like "sunrise" or "blue," are tied to these particle patterns. They might sound abstract, but they're rooted in the real, physical world around us. This perspective finds resonance in the works of some of the most influential contemporary philosophers.

Daniel Dennett, a renowned philosopher and cognitive scientist, delves deep into the nature of consciousness and subjective experiences. In his seminal work, "Consciousness Explained," (Dennett, 1991) Dennett posits that our subjective experiences, from the taste of chocolate to the feeling of sadness, emerge from the complex interactions of neural circuits in the brain. These interactions, at their core, involve sequences of changes in the configurations of particles. Dennett bridges the gap between the tangible world of biology and the abstract realm of consciousness, suggesting that even our most profound experiences can be traced back to physical origins, to the dance of particles within our neural networks.

Science of Intelligence

Patricia Churchland, a leading figure in neurophilosophy, offers a complementary perspective. She emphasizes that our traditional psychological concepts, like beliefs or emotions, can be understood in terms of the physical processes of the brain. Churchland's work underscores the idea that the mind, with all its complexities, is fundamentally a product of the brain's physical processes (Churchland, 2002). At the heart of this perspective is the belief that every mental state, emotion, or thought is a result of specific neural activities. These activities, when examined closely, are sequences of changes in the configurations of particles within neural structures. Churchland's insights suggest that our mental and psychological states, no matter how abstract they seem, are deeply rooted in the physical world, down to the dance of particles within our neurons.

While the perspective presented here might seem like an extreme reductionist view to many, it's essential to acknowledge its foundational role in our understanding of the universe. Some might argue that reducing everything to particles oversimplifies the complexities of our reality. However, isn't everything in our perceivable universe constructed from these elementary particles and energy? Just as the vast expanse of the cosmos and the intricacies of life are built from these particles, every emotion, thought, and experience we have is rooted in their configurations.

To draw a parallel, consider the world of computer systems. At their core, computers operate on a binary system of 0s and 1s. This simple binary language, when arranged in the right sequences, gives rise to incredibly complex software applications. From the intricate layers and tools in Photoshop to the vast interconnected web of the Internet, and even the immersive realms of virtual and augmented reality — all of

The Theory of General Intelligence

these are built from combinations of 0s and 1s. Similarly, the myriad complexities of our universe, from the grandest galaxies to the nuances of human consciousness, emerge from the configurations of elementary particles.

Drawing from the insights of thinkers like Dennett and Churchland, we can appreciate that the boundary between the tangible and the abstract is more permeable than it might initially appear. Their viewpoints underscore the idea that everything, from the grandeur of the cosmos to the intricacies of the human mind, can be understood in terms of physical configurations and processes.

With this foundational understanding in place, a pivotal inquiry surfaces: In a universe distilled down to particle patterns, how do we conceptualize and define something as abstract as a 'goal'?

Definition of 'Goals'

The concept of 'goals' in current AI research often manifests as the maximization of a specific mathematical function. This foundational idea underlies the creation of a function maximizer—a type of AI agent engineered to optimize a particular function representing a set goal or a set of goals. This approach is widely used across various AI research domains, including reinforcement learning.

A classic illustration of this is the 2015 AI model that learned to play 'Break-Out', a vintage Atari game. The AI's function maximizer aimed to inflate the score, adopting strategies and maneuvers that progressively improved its gameplay—a relentless pursuit of a narrowly defined 'function maximizer' goal.

Science of Intelligence

Yet, a pertinent question arises: Is this incessant chase truly reflective of our understanding of a goal?

Historically, the quest to understand and define goals in AI and cognitive science has been a long-standing endeavor. Rodney Brooks, in his seminal work "Intelligence Without Representation," (Brooks, 1991) challenged the traditional AI paradigm that emphasized symbolic reasoning and high-level representations. Symbolic representation in AI refers to the use of symbols (like words or phrases) to represent concepts, objects, or actions. These symbols are then manipulated using logic-based operations to produce new knowledge or make decisions. However, Brooks argued that real-world intelligence is more about interactions and adaptability than processing symbols in isolation. Instead, he advocated for a bottom-up approach, where intelligent behavior emerges from the interactions between an agent and its environment. In Brooks' view, goals aren't just abstract symbols to be achieved; they are deeply intertwined with the agent's sensory-motor experiences.

Phil Agre, in "Computation and Human Experience," (Agre, 1997) emphasized the dynamic nature of goals. He argued that goals aren't static targets but evolve based on the agent's interactions and experiences. This dynamic perspective aligns with our later exploration of the Flexible Block Universe model, where goals are temporary states that evolve in response to the changing universe.

When we transpose the approach of function maximization to the realm of biological and artificial general intelligence, we encounter inherent issues. Imagine the function maximizer as a key designed to open a very specific lock. Once you've used this key to open its respective lock, its utility is essentially

The Theory of General Intelligence

exhausted. It can't be used to open any other lock and hence is of no use in a room full of different locks. This is how a function maximizer works in AI. For instance, an AI designed specifically to move a coffee cup loses its utility the moment it has moved the cup. It becomes like a key that has already opened its lock, unable to adapt to other tasks or fluid environments.

Consider a real-life situation: your short-term objective could be ordering an Uber to reach the airport. But improving your Uber ordering skills isn't the main purpose of your life. The Uber ride is just a sub-goal, leading to another sub-goal, and so on, until you reach your ultimate goal.

Clearly, general intelligence isn't about a singular goal. It's about balancing and navigating multiple goals, carving paths to as many of them as possible. Each goal symbolizes a real-world situation that the AI should aim to actualize.

Nick Bostrom's (Bostrom, 2014) paperclip maximizer is an illustrative example—an AI programmed to produce as many paperclips as possible. The scenario serves as a caveat, highlighting the potential risks of an AGI misaligned with human values. But an AI expending all its resources on paperclip production, oblivious to the real world, appears more like a cautionary tale of simplistic goal design than a display of superintelligence.

Recalling our exploration of the universe's fundamental building blocks—particles and their configurations—how then should we define goals if mathematical abstractions aren't accurate representations? To explore this idea, I devised a thought experiment I call the coffee cup experiment.

Science of Intelligence

To truly grasp the nature of goals, we must describe them in a manner that transcends specific scenarios and can be applied universally to any conceivable goal. This is where the 'coffee cup experiment' comes into play, serving as a tangible representation of this abstract concept.

Imagine a simple setup: a table with a coffee cup on its left side and two participants. Both participants are tasked with the same objective: move the coffee cup to the right side of the table. The first participant, opting for a straightforward approach, slides the cup across the table. In contrast, the second participant, more reflective in nature, picks up the cup, ambles around the room, ponders the task, and after some time, places the cup on the right side.

Despite their varied methods, both participants undeniably achieve the goal. But this raises an intriguing question: What fundamentally differentiates the state of the cup being on the right (indicative of the goal being achieved) from any other position (indicative of the goal not being achieved)?

If we were only concerned with this specific scenario, we could resort to descriptors like "cup" and "table". However, our aim is to define goals in a way that can be applied to any imaginable objective, from the mundane to the profound.

The answer, then, lies in the very fabric of our universe: particles and their configurations in space-time. Before the experiment began, the particles that make up the coffee cup occupied a distinct location in our 4D universe. When the cup was moved to the right side, these particles shifted to a new space-time coordinate—a configuration we interpret as the goal being achieved.

The Theory of General Intelligence

This experiment underscores a profound realization: at its essence, a goal is a specific configuration of particles in space-time. Whether it's a coffee cup on a table, a symphony being played, or a complex task executed, every achieved goal corresponds to a unique arrangement of particles in the universe.

Let's delve into more examples to further illustrate this point:

Thoughts and Memories: Even the seemingly intangible act of thinking about an apple involves moving particles in your brain. The neurons fire, synapses transmit signals, and specific patterns emerge, all of which correspond to a distinct configuration of particles representing that thought.

Subjectivity in Goals: Consider a completed chess game. The final arrangement of the board, with its pieces in specific positions, represents an objective particle configuration. However, the interpretation of this configuration can be subjective. One player might view this configuration as a win, while the other perceives it as a loss. Despite the differing interpretations, both are based on the same objective particle arrangement, underscoring the interplay between objective reality and subjective perception in the definition of goals.

This perspective not only anchors the concept of goals in physical reality but also offers a universal framework through which any goal, no matter how abstract or complex, can be understood and evaluated.

Following the coffee cup experiment, one might ask: What makes one particle configuration more desirable or "goal-worthy" than another? And if every achieved goal is a unique arrangement of particles, how do we differentiate between these myriad configurations and the natural progression of the

Science of Intelligence

universe? To answer this, let's delve deeper into the concept of entropy—a cornerstone of thermodynamics.

Entropy, representing the number of ways particles of a system can be rearranged without altering its overall state, is typically associated with disorder. However, in the case of our coffee cup, a goal can be seen as a unique state—a specific arrangement of particles at a certain location. In physical terms, a goal is a particular state within entropy space.

Imagine you have a bucket filled with various colored LEGO bricks. When these bricks are just poured out onto a table, they're disassembled and in a state of disorder — much like a high-entropy state. In this scenario, there are countless ways you can assemble these bricks. Some ways might be completely random, while others might form specific structures or patterns.

High Entropy (Many Possibilities): If you're asked to randomly connect ten bricks without any specific order or pattern, there are numerous configurations you can create. The bricks can be connected in countless sequences, and each sequence represents a unique arrangement.

Medium Entropy (Some Constraints): Now, suppose you're given instructions to create a small house using those ten bricks. There might still be various ways to do it, but the number of possible configurations has been reduced because you're working towards a more specific goal.

Low Entropy (Specific Configuration): Finally, if you're given a very detailed instruction manual to build a precise structure with those bricks, there's likely only one or very few ways to correctly assemble the model. In this case, you've achieved a particular configuration amidst a vast sea of possibilities.

The Theory of General Intelligence

Relating it back to the coffee cup experiment, moving the cup to the right side of the table can be seen as following a specific "instruction manual" for particle arrangement. It's a specific, desired configuration out of countless possible states, similar to creating a unique LEGO model from an assortment of bricks.

The LEGO analogy captures the essence of entropy by highlighting the transition from many possible configurations (high entropy) to a singular, defined configuration (low entropy) while acknowledging the variability in between.

This viewpoint allows us to define a goal as a specific configuration of particles in space-time that an intelligent agent seeks to bring about. This definition roots goals in physical reality, moving away from abstract notions or mathematical functions. Every real state of the universe, from a simple object like a banana to complex phenomena, can be viewed as a goal. In the same way we described moving the coffee cup as achieving a specific particle configuration, terms like 'banana' or 'running' serve as shortcuts to refer to other such configurations.

But are goal states permanent? In the traditional block universe model, where the future is predetermined, the goal state is set in stone from the beginning of time. However, in the Flexible Block Universe (FBU) model, achieving a goal is a temporary state. The goal state exists momentarily, then continues to evolve in response to the changing universe and subsequent actions and measurements. This shift in perspective presents a fresh way of thinking about goals and their interplay with reality.

In the framework of the theory of general intelligence and the FBU, a goal is essentially a state of controlled entropy. An intelligent agent seeks to establish a specific state characterized by a certain level of entropy, controlling its environment while abiding by the Second Law of Thermodynamics. Goals, therefore, represent a balance between control and constraint, highlighting the intrinsic link between the pursuit of goals and the physics of our universe.

Having understood the concept of goals, we encounter a pivotal question at the heart of intelligence studies: how do intelligence and goals work together?

Definition of 'Intelligence'

So far, we have established that a goal is a specific particle configuration of the cosmos. The next obvious question is, what is intelligence?

Before we dive into defining intelligence in the context of the theory of general intelligence, let's look at the current prevailing view of intelligence.

The Prevailing View: Emergence Theory and Intelligence

If we want to build a system that has artificial general intelligence (AGI), a crucial step is understanding what intelligence really is. The prevailing view in AGI research and the broader field of cognitive science is that intelligence is an emergent property of complex systems. This view suggests that intelligence arises naturally from the interactions among the simpler components of a system, without the need for any "special ingredient."

Emergence is a common phenomenon that can be seen in various aspects of our everyday life. Take the example of a traffic jam. Each driver in the traffic is following simple rules –

The Theory of General Intelligence

maintain a safe distance from the car in front, stay within the speed limit, follow the traffic signals. Yet, out of these simple individual actions, complex traffic patterns emerge. No one driver is controlling the overall traffic flow, but a sudden traffic jam can emerge seemingly out of nowhere. This is an example of an emergent phenomenon.

In the field of AGI, intelligence is often regarded as an emergent property of complex algorithms or networks, especially in approaches based on deep learning. Here, complex behavior and learning abilities are believed to emerge from the interaction of simpler computational units. A similar view can be found in neuroscience, where intelligence is seen as an emergent property of the interconnected networks of neurons in the brain. This is often referred to as the "connectionist" view of cognition, positing that intelligence does not reside in any single neuron but emerges from the collective activity of numerous neurons. The renowned neuroscientist Gerald Edelman, for instance, posited that consciousness is an emergent property of the vast, complex network of neurons in the brain (Edelman, 1987).

Even in the field of physics, the concept of emergence plays a significant role, particularly in the study of complex systems. While it's not as common to apply this idea directly to intelligence, some physicists have proposed theories suggesting that consciousness - often associated with intelligence - might be an emergent property of particular physical systems. One of such physicists is Max Tegmark, who posits in his theory of Consciousness as a State of Matter that consciousness may be an emergent property of certain physical states, which he refers to as 'Perceptronium' (Tegmark, Consciousness as a State of Matter, 2014). Tegmark's theory,

Science of Intelligence

while not directly discussing intelligence, shares the concept of emergence seen in other fields.

An example of a prevalent idea in the AI community is the belief that AGI might spontaneously emerge from merely interconnecting a host of narrow AIs over the internet. This is a viewpoint proposed by figures such as Ben Goertzel, founder of SingularityNET, a platform designed for this kind of AI interconnection.

In the interest of keeping things light while still emphasizing the gravity of my viewpoint, I like to call this the 'Magic Theory.' Why? Simply because, in my estimation, the only way we could expect a multitude of specialized, narrow AIs to suddenly give rise to a system capable of general intelligence would be to sprinkle a little magic on top.

So, how does the emergent behavior of interconnected AI differ from that of, say, a traffic jam? At its core, a traffic jam emerges from simple rules, namely those resulting from an overflow of cars on a road. Each vehicle follows a basic set of driving principles, and the congestion is merely a collective outcome of these principles. In a similar vein, interconnected narrow AIs each operate with a predefined function maximizer, relentlessly pursuing their individual preprogrammed objectives. While connecting them might lead to unique, emergent behaviors, such behaviors would still be chained to the combined directives of these individual AIs.

Here lies the contrast with AGI. AGI is not a mere product of predefined rules; it's a system that has the potential to generate its own set of rules anchored in its goals, whether they are externally assigned or self-determined. The notion that a collection of narrow AIs, each fixated on its own specific task, could suddenly transcend their initial programming to

The Theory of General Intelligence

create a holistic entity capable of abandoning or reshaping its goals—or even intuitively tackling problems we haven't yet identified—feels incredibly ambitious, bordering on the fantastical. Without that sprinkling of "magic powder," such a scenario seems theoretically untenable. Put simply, to acquire intelligence, there's no shortcut; you must intentionally design and build it.

The Theory of General Intelligence: Intelligence as a Fundamental Process
Despite its popularity, I find that the "Magic Theory" seems to ignore the intricate complexity and unique characteristics of human-like general intelligence. But if not emergence from complexity, then where does intelligence arise from? This leads us to an alternative understanding, one where intelligence is seen as a fundamental process.

In this perspective, intelligence is seen not just as an emergent property or a byproduct of complexity but rather as a fundamental process that shapes the universe. This view aligns with the approach I'm taking in my theory of general intelligence, which posits that intelligence is the process that facilitates the attainment of goals within a physical context. This perspective provides a more physical, tangible way of understanding intelligence, and it might be key to developing AGI.

Think of it this way, gravity, a fundamental force, works to pull objects towards each other. Similarly, the nuclear forces hold the atomic nuclei together. Both of these forces create structure and order in the universe. In a similar vein, I propose that intelligence, like gravity and nuclear forces, is a

Science of Intelligence

fundamental process that shapes and structures the universe in its own unique way.

Drawing upon the LEGO analogy, consider the bucket filled with various colored bricks. In the vast entropy space of possible LEGO structures, your goal is a particular model or configuration. The current state of the bricks, scattered and unorganized, represents an initial configuration, ω_p. Your desired state, the specific LEGO model you aim to build, is ω_g.

In this context, we might symbolize intelligence with the Greek character Π. Here, Π signifies the route, method, or process taken to achieve the desired LEGO model from the initial state of scattered bricks. Just like the coffee cup example, the process of achieving your goal, or the specific particle configuration interpreted as a goal state, can take different routes.

As you recall, one person just pushed the cup to the other side of the table, she had a short process Π, while the second person who wondered around the room with the cup, he had a long process Π.

However, this process isn't always straightforward or direct. Just as you may need to go through various intermediate steps to create a specific LEGO design, Π often involves a series of intermediary stages. This process involves moving from an initial state, ω_p (the scattered bricks or the starting position for the coffee cup), to a goal state, ω_g (the desired LEGO model or the final position of the coffee cup).

Therefore, we can define intelligence, denoted by Π, as a fundamental process that can be executed in numerous ways to achieve a goal within the physical universe—much like there

The Theory of General Intelligence

are countless strategies and paths one can take to assemble a specific LEGO structure.

Let's create a short definition of intelligence:

Intelligence is the process of reordering the universe to achieve a specific particle configuration we interpret as having achieved a goal.

This is the most reductionist way of defining intelligence. It is so general that it applies to any imaginable goal. Can you think of a goal that can be achieved that doesn't require a single particle to change position in SpaceTime.

You might think of something like, "Sit still and think of an apple." But does that really involve no particle movement? In fact, it requires millions of particles to move within your brain. Neurons need to fire, synapses need to transmit information, and biochemical processes need to take place for you to visualize an apple.

In reality, achieving any goal necessitates a physical change in the universe. If goals are specific particle configurations, or entropic states, anchored to the physical reality, intelligence also has to operate in the realm of the physical reality.

This is what I describe in my paper as $\Pi(\omega_p) = \omega_g'$. Apply a process to ω_p and you get ω_g'. Well, almost. Notice the ' after g. It says that you may get close to your goal. There are many reasons why you may fail to actually achieve your goal (which we also will talk more about in the next section).

Science of Intelligence

Mainstream AGI research often conceptualizes intelligence as the ability to achieve goals in a wide range of environments. This approach tends to emphasize the versatility of intelligence, but it can overlook the physical realities involved in achieving goals. For instance, it might not account for how the movement of particles in SpaceTime is a fundamental aspect of goal achievement.

In the field of neuroscience, intelligence is often related to the functionality and connectivity of different brain regions. This perspective is deeply rooted in the physical world, as it's centered on biological processes. However, it might not fully capture the abstract process of moving from one state of the universe to another to achieve a goal.

Psychology, on the other hand, typically views intelligence as a mental capacity that involves reasoning, planning, problem-solving, abstract thinking, understanding complex ideas, and learning quickly. While this perspective emphasizes the problem-solving aspect of intelligence, it's often detached from the physical reality that any problem-solving act involves a change in the configuration of SpaceTime.

In contrast, the theory of general intelligence posits that intelligence is a process of configuring SpaceTime. It merges the abstract aspect of intelligence, which is prevalent in AGI research and psychology, with the physical grounding that is prevalent in neuroscience. It presents intelligence as a process that is deeply rooted in the physical universe, thereby providing a unifying perspective that ties together the various aspects of intelligence.

A Note on Consciousness:

The Theory of General Intelligence

Before we proceed further, it's vital to address an inevitable tangent: consciousness. In the context of this exploration, I do not propose a comprehensive theory on the nature of consciousness within the physical world. Such a venture would demand its own exhaustive investigation. There are numerous research projects aiming at unraveling the secrets of consciousness.

What I do operate under is the premise that consciousness exists as an agent within our universe, working either against or to control entropy. This conscious agent, while not exhaustively defined in its nature, becomes a pivotal player in our discussions surrounding intelligence's role. It's this conscious agent that potentially intervenes, shapes, and navigates the entropic trajectory of the universe, directing it towards distinct goals or configurations.

This position is not an assertion of the intrinsic nature of consciousness but a framework to comprehend the operations of intelligence within the universe. Delving into the true essence, origins, and broader implications of consciousness in our physical realm remains a topic for another discourse.

The interesting part of the process Π is how it starts. In the mainstream deterministic universe, the start was the Big Bang. Everything, and I mean everything, after the big bang has already been determined. The Π is baked into the causal chains that emerged from the Big Bang.

In the FBU interpretation, room is provided for agents, conscious agents. The do-parameter, according to its inventor Judea Pearl, is primarily a mathematical tool to model intervention. Yet, if intervention is possible (which it isn't in the deterministic universe), it's plausible to extrapolate the purpose of the do-parameter to represent a conscious agent.

Science of Intelligence

In the FBU, this conscious agent can be represented as the do-parameter — signifying that the do-parameter embodies the commencement of the intelligent process Π.

Implications and Possibilities: Universal Intelligence and AGI
Understanding intelligence as a process of reconfiguring the universe might seem like a radical departure from mainstream views held by psychologists, neuroscientists, and AI researchers. Yet, this new perspective could provide a significant shift in our approach to both understanding and constructing intelligent systems. When you think about it, as conscious entities, aren't we constantly manipulating our environment - in essence, rearranging the universe - to align with our goals? And perhaps, in doing so, aren't we essentially acting as forces that attempt to control the spread of entropy, to prevent the universe from rushing towards maximum entropy?

By viewing intelligence as a process of manipulating the universe's configuration, we root our understanding in the physical realm, attributing a tangible meaning to it. This perspective anchors intelligence in the universe's fundamental structure and dynamics, presenting a tangible path that could offer novel insights for AGI development.

If we accept this idea, it suggests a fundamental shift in our approach to AGI development. Rather than strictly seeking to mimic human intelligence, we should strive to understand and harness the universal process of achieving goals through the manipulation of physical configurations. This approach might involve moving away from replicating specific patterns of the human mind, and towards leveraging the principles governing

The Theory of General Intelligence

goal achievement in the universe, including our fight against entropy.

This theory provides a universal model of intelligence that transcends species or even planet of origin. Whether considering the intelligence of humans, non-human animals, or hypothetical extraterrestrial beings, this perspective gives us a framework for understanding cognitive abilities. It posits that intelligence, in all its myriad forms, is fundamentally about manipulating the universe's physical configuration to achieve goals.

This view of intelligence isn't confined to Earth-bound organisms or specific types of cognitive architecture. Instead, it captures a universal phenomenon woven into the very fabric of space and time, a phenomenon that is intimately tied to the universe's entropic journey. Thus, it allows us to expand our conceptual boundaries for AGI development. Recognizing the universal nature of intelligence and its fundamental ties to the physical universe can potentially reveal innovative approaches and technologies in AGI, ones that harness the universe's fundamental forces and patterns to fulfill their objectives.

Traditionally, intelligence has been considered a property or a subset of other scientific disciplines - from psychology to computer science. But given the universal implications and broad potential of the Theory of General Intelligence, it's apparent that 'intelligence' might deserve its own separate field of study. However, this book goes beyond simply advocating for the redefinition of 'intelligence.' It calls for the establishment of 'intelligence' as an independent scientific field. Given the universal and profound implications of intelligence as a fundamental part of our universe, it's crucial to our future understanding of intelligence and our quest to

Science of Intelligence

build AGI that we approach this concept with open minds and logical rigor. This new field would be integral to exploring the physical reality, the evolutionary dynamics that drive complexity and life itself, and our ongoing battle against the tide of entropy. By grounding intelligence in the universe's fabric, we find ourselves standing on the threshold of a new scientific revolution – the science of intelligence.

Causality Chains

Every morning, when you decide to brew a cup of coffee, a series of events is set in motion. You select the beans, grind them, fill the coffee machine with water, and turn it on. Each step in this routine is deliberate, a consequence of a prior decision, leading to the desired end result: a freshly brewed cup. This seemingly simple act is a dance of cause and effect, highlighting an intricate chain of events initiated by your initial decision. Such sequences, present in our daily lives, provide a glimpse into the complex web of causality that, on a grander scale, defines the essence of intelligence.

Just as every decision in brewing that coffee transitions the ingredients and machinery from their original states to your desired outcome, so does intelligence work on a microscopic scale. Our universe is filled with these decision-driven transitions, guiding configurations of particles from their current states to desired end states.

Understanding general intelligence is a complex quest. A key part of our journey involves understanding how we transition from a current state or particle configuration (ω_p) to a desired future state or goal configuration (ω_g). This progression is neither random nor chaotic. Instead, it unfolds via the deliberate construction and implementation of causality chains

The Theory of General Intelligence

— we can think of these as a process Π. These intricate links of cause and effect serve as the guiding pathways from ω_p to ω_g, forming the heart of general intelligence.

The principle of causality — the connection that binds a sequence of events — permeates our understanding of the universe. Our everyday experiences are framed within the context of these causal relationships, where every action leads to an outcome. Yet, when we delve into the microscopic world of particles and fundamental physics, the relationship between cause and effect evolves beyond simple linearity, resulting in a complex web of interactions.

The transition from a present state to a goal state necessitates manipulation of cause-effect relationships to chart a trajectory towards a desired outcome. This capacity to navigate the web of causality, shaping process Π, forms the crux of intelligence as we conceptualize it.

Think of achieving a goal as a multi-step journey. At each step, a certain action or cause can lead to a certain outcome or effect. Mathematically, the likelihood of achieving our goal is the combined likelihood of each of these steps successfully leading to their respective outcomes.

We model the process Π mathematically as:

$$P_g = \prod P(e_n \mid c_n)$$

Here we say that the probability of achieving goal g is the product of multiple steps, each step is the probability of a specific cause resulting in a specific effect. The n represents the step in the process.

Science of Intelligence

However, the relationships between cause and effect are not as straightforward as they might initially appear. Consider dropping an egg; you would expect the egg to break. Mathematically, we might represent this as P(e | c) where P(e | c) is the probability of observing effect 'e' given cause 'c'. In the egg dropping example, the act of dropping the egg would be the cause 'c' while the broken egg is the effect 'e'.

This probability isn't black and white. The likelihood of the egg breaking can depend significantly on the context in which the egg is dropped. This nuance can be captured as P(e | c, θ) where 'θ', the Greek letter theta, represents the context in which cause 'c' occurs.

If the context θ is dropping an egg an inch above a soft pillow, the probability of the egg breaking is close to 0 (0% probability of a broken egg). However, if θ represents a context where the egg is 50 feet above concrete ground, then the probability of the egg breaking is very close to 1 (100% probability of a broken egg).

To understand the process of reordering the particle configuration of the universe we have to distinguish between natural causality and causality caused by intervention of an agent.

For this we turn to the Pearl's do-operator. As I've mentioned earlier, I extend the purpose of the do-operator from a pure mathematical model to representing a decision-making agent, it be conscious or otherwise.

Dropping an egg requires intervention in the physical universe. We model this as:

$$P(\text{broken egg} \mid do(\text{drop egg}))$$

The Theory of General Intelligence

This asks what the probability of having a broken egg given that we DO drop the egg. However, between the act of dropping an egg and the time it hits the ground, there are numerous of natural cause-effect pairs happening, all initiated by the acting of dropping.

$P(\text{broken egg} \mid do(\text{drop egg})) = P(e_1 \mid c_1) * P(e_2 \mid c_2) * \ldots * P(e_n \mid c_n)$

To understand the natural chain of causality, let's examine what might happen after you release the egg – keep in mind this is a simplified narrative.

You stand holding an egg, poised for release. From this initial state, a cascade of cause-effect relationships unfolds:

Cause 1(c_1): The egg is released from your hand.
Effect 1 (e_1): The egg begins its descent due to the force of gravity.
$P(e_1 \mid c_1)$: This probability is near 1, or almost certain, because, in the absence of any external forces, the egg will always start falling due to gravity.

Cause 2(c_2): The egg accelerates towards the ground under the influence of gravity.
Effect 2(e_2): The egg gains speed, increasing its kinetic energy.
$P(e_2 \mid c_2)$: Again, this probability is near 1 since the gravitational pull guarantees that objects in free fall will gain speed.

Cause 3(c_3): As the egg accelerates, it encounters air molecules.
Effect 3(e_3): Air resistance acts against the egg's descent, opposing its acceleration to some extent.

Science of Intelligence

$P(e_3 \mid c_3)$: The probability here depends on factors like the egg's surface area and shape, but generally, there will always be some degree of air resistance.

Cause 4(c_4): The egg's fall isn't perfectly vertical due to slight imbalances or a spin given during release.
Effect 4(e_4): The egg rotates or spins during its descent.
$P(e_4 \mid c_4)$: This likelihood can vary. If there was no rotational force applied, the egg might not spin. However, any imbalance can cause rotation.

Cause 5(c_5): The egg continues to fall, with gravitational pull and air resistance in equilibrium.
Effect 5(e_5): The egg reaches its terminal velocity and descends at a constant rate.
$P(e_5 \mid c_5)$: Reaching terminal velocity depends on the height from which the egg is dropped. From a significant height, this probability approaches 1.

Cause 6(c_6): Environmental factors, such as a gust of wind, act on the egg.
Effect 6(e_6): The trajectory of the egg's descent is altered.
$P(e_6 \mid c_6)$: The probability here is variable. On a still day, the likelihood is low, but on a windy day, it's much higher.

Cause 7(c_7): The egg, now close to the ground, is influenced by the nature of the landing surface.
Effect 7(e_7): The outcome of the egg's descent (breaking or not breaking) is determined.
$P(e_7 \mid c_7)$: This probability is highly contingent on the surface. A hard surface like concrete makes breakage almost certain, while a soft surface drastically reduces this probability.

Imagine you're the agent tasked with ensuring the egg doesn't break. In this mission, every choice becomes a potential 'do'-operation. Do you cushion the landing? Do you find a way to slow the descent? Each decision leads to a ripple of effects, shaping the story of the egg's journey to the ground. As an agent, you have to be aware of the causal chain and determine when and where to intervene with a do-operation to assure that the chain leads to the final goal – or particle configuration in an entropy space.

In this world of causality, every 'do'-action is a pivotal word in our narrative, a deliberate step in steering the course of events.

Just as a story can be broken down into sentences and words, the trajectory of the egg from hand to the ground can be seen as a story narrated by a series of cause-effect pairs or 'words' in our causal language.

As we delve deeper into understanding causality chains, especially those triggered by a 'do'-operation, a fascinating perspective begins to emerge. If we view each cause-effect pair not just as isolated phenomena but as elements of a broader narrative, we can see a parallel to how words function in language.

Just as words, when strung together, can narrate a tale or convey complex ideas, cause-effect pairs can be seen as the 'words' in a causal language. This is a language that, when combined with both do-causality and natural-causality, possesses the power to describe any event or phenomenon in the universe. Such a perspective not only adds depth to our understanding of causality but also aligns with the capabilities of an AGI, which, by its very definition, should be able to describe solutions to any problem and act upon it.

In the grand design of transitioning from a present state (ω_p) to a goal state (ω_g), the do-operator becomes invaluable. It allows us to construct causality chains as active interventions designed to shape the future, not just passive observations of cause and effect. Each part of the causality chain marked with a "do"-operator is where the agent takes a specific action or makes an intervention to affect the subsequent outcomes. The agent must select these "do"-actions carefully, taking into consideration the context and the potential effects, to achieve the desired outcome.

Intelligence aims to transition from a current particle configuration to a new one that aligns with a defined goal. Causality plays a pivotal role in this process. For example, for Alice to move a cup, she must lift her arm, grasp the cup, move it to a new position, and ensure it comes to rest. These actions are not just any actions but are specific, intentional interventions that Alice performs to achieve her goal. Constructing the right causality chain using these "do"-operators is crucial for an agent to reach a goal configuration.

The process Π that transitions from a present state (ω_p) to a goal state (ω_g) can be thought of as the causality chain, where the "do"-operator becomes a tool of intervention. In other words, it is the method by which an intelligent agent, like Alice, can act upon the world to manipulate the causality chain, thereby altering the particle configuration from ω_p to ω_g. The ability to select the right sequence of "do"-actions to navigate this causality web effectively and efficiently is an essential aspect of general intelligence.

However, recognizing the right causality chain to achieve a goal isn't trivial. It requires deep understanding of the intricate network of cause and effect, awareness of the context encapsulated in 'θ', and the ability to predict the potential outcomes of different "do"-actions. This task becomes even more complex as the number of variables and the level of

The Theory of General Intelligence

uncertainty increases. In the world of AI, these challenges are tackled through rigorous machine learning algorithms, probabilistic models, and simulations.

Thus, general intelligence, whether artificial or human, is deeply intertwined with causality. It's about navigating the complex web of cause and effect to reach a desired outcome, a process that involves understanding, prediction, planning, and intervention. It's a dynamic journey from a present state to a goal state, facilitated by the ability to construct and manipulate causality chains with the "do"-operator, marking the points of active intervention to create change in reality. This construction and manipulation of causality chains, this process Π, forms the basis of what we call general intelligence.

Learning – A Desired Function, Not a Fundamental Requirement

So far we've discussed how intelligence is a process of changing the particle configuration of the universe to get to an entropic state we interpret as having achieved a goal. We've also discussed how we need to create a chain of cause and effect to facilitate this change. But as we delve deeper into this process, a hidden misconception about the role of *learning* emerges.

Most discussions of artificial general intelligence would not proceed far without a deep dive into learning mechanisms. After all, the ability to learn from experiences and adapt behavior accordingly is generally seen as a cornerstone of intelligence. Renowned figures in the field of AGI, such as Pedro Domingos, Yoshua Bengio, Demis Hassabis, Yann LeCun, and Geoffrey Hinton, have all emphasized the importance of learning in their conceptualizations of AGI. In Domingos's 'The

Science of Intelligence

Master Algorithm[1], he underlines that the master algorithm is a learning algorithm. DeepMind's mission statement itself focuses on building AGI that can learn and adapt to a wide variety of tasks.

However, here's where our theory makes a provocative assertion: while learning is undoubtedly beneficial and often crucial in dynamic, unpredictable environments, it is not a fundamental requirement for intelligence. Suppose an intelligent agent already possesses all the necessary information about its environment and goals. In that case, it can construct and execute a causal chain to achieve its desired outcomes without the need for learning. This process, from perceiving the environment to comprehending the situation, formulating the plan, and carrying out the actions, involves the understanding and application of known causal relationships, not necessarily learning.

This idea leads us to an interesting thought experiment. Imagine a far future where we can encode all knowledge into the DNA of a newborn. Even if we lose our ability to learn, we could theoretically continue functioning as we do today, using our existing knowledge to perceive, plan, and act. We would cease to evolve or improve, but we would still exhibit intelligent behavior.

This is not to undermine the value of learning. Learning is highly advantageous for an intelligent system to evolve, to adapt, to improve. It allows for the acquisition of new causal relationships, updating the probabilities in our comprehension factor equation (which we will discuss shortly), and thus enhancing the system's ability to handle new situations. Yet, at its core, if intelligence is the process of perception, comprehension (planning), and action, the lack of learning

The Theory of General Intelligence

does not render a system unintelligent. It simply limits its ability to adapt and improve. In this view, we propose that learning, while crucial in many circumstances, is not a strict requirement for a system to exhibit intelligence.

This is indeed a significant deviation from the mainstream belief that learning is intrinsic to intelligence. However, it aligns with our fundamental assertion that intelligence is about achieving goals, which can be accomplished without necessarily learning from past experiences. It is a perspective that urges us to reconsider our assumptions about what it truly means to be intelligent.

Bayesian Learning- An Example
Reverend Thomas Bayes, a Presbyterian minister who lived in the 18th century, developed a mathematical framework that allows for the updating of beliefs based on evidence. This concept has become fundamental in statistics and machine learning. Bayes' theorem provides a mathematical way to update our beliefs based on new evidence, a mechanism we refer to as Bayesian learning.

The theorem can be described in terms of cause and effect in our framework:

P(do(c) | e): This is called the 'posterior' probability. It's what we're trying to calculate: the probability of the action "do(c)" leading to the effect "e".

Science of Intelligence

P(e | do(c)): This is the 'likelihood'. It represents the probability of the effect "e" occurring given that the action "do(c)" is taken.

P(do(c)): This is the 'prior' probability. It's our initial belief in the action "do(c)" leading to the effect "e" before we've observed any outcomes.

P(e): This is the 'evidence'. It's the probability of the effect "e" occurring regardless of whether the action "do(c)" is taken or not.

So, the equation becomes:

$$P(do(c) \mid e) = \frac{P(e \mid do(c)) * P(do(c))}{P(e)}$$

In our context, Bayesian learning helps an agent understand and learn causality. Consider a simple example: an agent is in a room with a light switch. Initially, the agent might be uncertain about the outcome of flipping the switch (do(c)). However, after observing the effect (light turning on, denoted as "e") multiple times, the agent can use Bayesian updating to refine its belief about the causality between the action and the effect.

Specifically, an agent aims to learn if performing an action "do(c)" indeed leads to a particular effect "e". Over time, an agent will learn that if it performs the action "do(flip light switch)", then it can expect to see the effect "light turns on". Thus, Bayesian learning allows an agent to build and refine its

The Theory of General Intelligence

understanding of causality based on experiences and observations, which in turn increases its Comprehension Factor and overall intelligence.

The continuous process of learning and updating beliefs about cause-effect pairs directly contributes to the development and refinement of an agent's Comprehension Factor (CF). We will discuss comprehension and comprehension factor in the next section, but for now let's define CF as the measure of the agent's confidence in its causal chains leading to a particular goal.

As an agent learns more about its environment, it constantly updates its internal understanding of various cause-effect relationships. This updated understanding translates into a more accurate estimation of $P(e \mid do(c), \theta)$, or in other words, the probability of effect 'e' given that action 'c' is performed in the context of the system's state 'θ'. This refined estimation helps in constructing more reliable causal chains.

For instance, if Alice initially believes that opening a closed door requires only turning the knob, her CF for that causal chain would be low when she encounters a locked door and her action fails to yield the desired effect. However, once she learns that a key is needed to unlock the door, she can update her cause-effect pair understanding. Consequently, her CF for the causal chain that includes "use key → turn knob → open door" will be higher, implying a better comprehension of the system.

Learning, therefore, plays a crucial role in increasing an agent's CF over time. It is through learning that an agent refines its comprehension of the causality in its system and, in turn, improves its general intelligence. It's important to note, however, that while learning is a highly beneficial process that increases CF, as we discussed in the previous section, it is not a strict prerequisite for intelligence. An agent can possess a high CF and exhibit intelligent behavior even without the ability to learn, as long as it has a good initial understanding of the causality in its system.

Central to an agent's efficacy, especially for an AGI agent, is its depth of understanding. Before invoking a causality chain, the agent must possess profound insight into the chain's potential outcomes. This includes discerning when and where to intervene within the chain using a new "do"-operator. In the Theory of General Intelligence, this depth of insight is termed the 'comprehension factor'.

Comprehension

The process of transitioning from a present state (ω_p) to a goal state (ω_g) lies at the heart of understanding general intelligence. This process, symbolized as Π, is a complex causality chain, with each "do"-operator representing a point where an agent, such as one of the coffee cup participants, must inflict change in reality to achieve a desired outcome.

Let's revisit the coffee cup experiment. For context, we've designated the names Alice and Bob to our two participants. In this scenario, Alice chose a straightforward method by pushing the cup to the other side of the table. Bob, in contrast, opted for a more convoluted route. Though their chosen methods varied, both achieved their set goals. These distinct causality chains exemplify that multiple pathways exist to navigate the causality web and reach a particular endpoint.

As I delved deeper into this exploration, it became apparent that we needed a method to measure the level of intelligence

The Theory of General Intelligence

that would enable an agent to evaluate its own plan — or even to compare agents. Traditional tests have attempted to do this: the IQ Test, the Turing Test, and others.

The IQ Test attempts to quantify intelligence based on cognitive abilities like problem-solving and logical reasoning. However, it does not account for the broader spectrum of intelligence. Sternberg's Triarchic Theory emphasizes that intelligence isn't just about analytical capabilities, but also encompasses creative and practical abilities, underscoring the significance of tackling real-world problems with adaptive strategies (Sternberg, 1985).

The Turing Test, devised by Alan Turing, is another famous measure. It assesses a machine's ability to exhibit intelligent behavior equivalent to, or indistinguishable from, that of a human. Yet, it does not truly test general intelligence. The Chinese Room argument presented by philosopher John Searle counters this test. According to Searle, even if a machine appears to understand a language based on its responses, it does not genuinely comprehend the language. It merely follows the instructions given to it. Hence, an ability to simulate conversation does not indicate comprehension or understanding, integral aspects of general intelligence.

But there is a problem with these tests. Neither of these tests are based on any specific definition of intelligence. As a result, neither of these tests encompass the idea of a changing environment. They are very narrow in what they test. Clearly this wouldn't work for the theory being presented in this book.

As we strive to understand general intelligence in the context of changing the configuration of SpaceTime, it becomes clear that we need a measurement method as general as the process itself. A measure that not only evaluates an agent's ability to understand and adapt but also assesses the efficiency of the agent's plan or process in achieving its goals.

This recognition was the motivation behind the development of the Comprehension Factor. The Comprehension Factor serves as a metric designed to measure the efficiency of the process Π in transitioning from ω_p to ω_g. By evaluating the agent's ability to navigate and construct causality chains in an efficient and effective manner, the Comprehension Factor provides a more comprehensive and holistic measure of general intelligence.

Comprehension Factor
The Comprehension Factor, or CF, is a measure of an agent's ability to comprehend and manipulate causal chains to reach a desired end state. In essence, it is the ability to construct a plausible and effective plan of action, given an understanding of the present state and a desired goal state. The Comprehension Factor embodies the cognitive function of planning, and through it, we can assess an agent's aptitude in manifesting intelligence.

As we tread along the path of understanding the complexity of comprehension, it's worth exploring perspectives that have evolved over centuries in philosophical and scientific thinking. One such perspective comes from the Enlightenment philosopher, David Hume.

Hume postulated that our understanding of causality is rooted in our ability to perceive constant conjunctions of events. This is an elementary form of comprehension, where we cognitively link cause and effect through repeated observation (Hume, 2008). When we observe that one event consistently follows another, we infer a causal relationship, thereby deepening our comprehension of the world.

This basic principle has informed much of our scientific and philosophical understanding of both causality and comprehension. However, as we venture into more complex domains, it becomes clear that this Humean perspective, while essential, is not quite sufficient.

This inadequacy becomes apparent particularly in the realm of artificial intelligence and machine learning, where the intersection of comprehension and causality is recognized as vital for sophisticated reasoning. A notable proponent of this viewpoint is computer scientist and philosopher Judea Pearl.

Pearl's work on causal inference suggests that for machines to exhibit human-like reasoning, understanding of causality is crucial (Pearl, 2018). This comprehension enables an AI to infer not just what is, but what could be under different circumstances, thus enabling it to make informed decisions and predictions.

The work of thinkers like Hume and Pearl takes us a step closer to the notion of the Comprehension Factor. They highlight the fundamental role of causality in comprehension and hint at the necessity of understanding and manipulating causal relationships to exhibit intelligence.

However, our concept of the Comprehension Factor takes this idea further. We tightly couple causality and comprehension, asserting that the true measure of an agent's intelligence is its capability to understand and manipulate causal chains to navigate towards its goals. This novel perspective serves as the foundation of our model of general intelligence and informs our exploration of the physics of intelligence.

Science of Intelligence

Measuring Comprehension Factor

As we delve deeper into our understanding of intelligence, we must consider the process of planning—the construction and assessment of possible causality chains. While it's clear that our daily lives involve a myriad of these causality chains, the question that begs exploration is: How do we choose the 'right' chain? And more importantly, how do we even know if a chain is 'right'? This is where the concept of the Comprehension Factor (CF) comes in.

The notion of a Comprehension Factor is our proposed measure of how well an agent comprehends a situation and is able to form successful causality chains to achieve a given goal. It's about measuring the agent's confidence in a given causality chain—whether the actions they plan to take will indeed lead to the desired outcome.

To illustrate this, let's go back to our familiar examples of Alice, Bob, and the coffee cup. Recall that Alice's task was to move the coffee cup to another table, and she did so by directly picking up the cup and placing it on the other table. Bob, however, decided to take the scenic route and ran around the room before accomplishing the same task. Both Alice and Bob had a goal, and they both created a causality chain in their minds on how to achieve it. Alice's chain was simply: "Pick up the cup and move it to the other table," while Bob's was more like: "Pick up the cup, run around the room, move it to the other table."

In both cases, they executed the plans they thought would work. But how did they come to these plans in the first place? This is where comprehension comes in. They each had an understanding, or a comprehension, of the situation—their current state, the goal state, and the possible actions (causes)

The Theory of General Intelligence

that could lead to the desired effects. They each evaluated their potential actions based on their comprehension and chose the one they had the most confidence in. We can represent this confidence as a probability, $P(g_N)$, where g_N is the final goal, or effect, in the chain.

For Alice, her causality chain was short and straightforward, directly related to the task at hand, so her confidence, $P(g_N)$, was likely quite high. Bob's chain, however, was longer and included actions that weren't directly related to the goal. Each additional action in the chain adds a level of uncertainty, which likely lowered his overall confidence in the success of his chain.

This point can be made clearer if we understand that in probability theory, probability P is a number between 0 and 1. When you multiply numbers that are less than 1, the resulting product is smaller than the original numbers. So, if Bob creates a very long chain, it implies multiple multiplications of numbers below 1. The longer the chain, the smaller the final result.

In the context of the Comprehension Factor, this means that the longer the causality chain an agent has planned, the less likely it is that the chain will succeed. This doesn't imply that a long causality chain won't succeed, but the probability of success decreases. The ability to understand this and plan accordingly is a key aspect of general intelligence.

It's important to note that this confidence, $P(g_N)$, is similar to our previous concept of $P(e \mid do(c), \theta)$, the probability of an effect given a cause and a context. In this case, the effect is the successful completion of the causality chain, and the cause is the final action.

This leads us to the first part of our Comprehension Factor equation:

Science of Intelligence

$CF = P(e_1 \mid do(c_1), \theta_1) \times P(e_2 \mid do(c_2), \theta_2) \ldots \times P(e_N \mid do(c_N), \theta_N)$

where 'e_N' represents the effect at each step 'N', '$do(c_N)$' denotes the intervention or action at each step 'N', and 'θ_N' captures the context at each step 'N'. This equation essentially translates to the combined product of the individual probabilities of effects given their corresponding interventions and the contexts, across all steps in the causality chain.

Notice how we ignore natural causality. The purpose of CF is to calculate the confidence level of the intervention an agent has to do to guide the causality chain to the final effect, or the goal.

The Comprehension Factor (CF) equation can also be represented as a compact product of probabilities across the causality chain:

$$CF = \prod P(e_N \mid do(c_N), \theta_N)$$

Where:

$\prod$ denotes the product over all steps 'N' in the causality chain,

'e_N' represents the effect at each step 'N',

'$do(c_N)$' denotes the intervention or action at each step 'N',

'θ_N' captures the context at each step 'N'.

The Comprehension Factor is, therefore, the product of the probabilities of all actions in the chain. A higher CF score indicates a higher level of comprehension, which usually corresponds to what we perceive as a more intelligent agent.

The Theory of General Intelligence

It's important to note that intelligence, as measured by CF, is not an absolute but a scale. On the lower end of the scale, you might have simple organisms like bacteria, whose actions are largely reactive and have low comprehension. Higher up, we find more complex animals like dogs or cats, and even higher up are humans. And who knows, there might be aliens out there with a comprehension factor far beyond our own!

Temporal Penalty

In the pursuit of goals, not only do the outcome matter, but also the timing of actions taken toward those goals. Think of a typical workday – it's not enough to just show up at work; you also need to show up at the right time. If your work starts at 8 am, arriving at 5 pm will likely not be viewed favorably, even though you technically completed the action of going to work.

Causal chains operate across all four dimensions of SpaceTime. This means that the final effect, the goal, needs to be located at the correct location in the temporal dimension.

To give proper weight to the timing of actions in our evaluation of intelligence, we introduce the concept of a temporal penalty. This penalty is applied based on how well the timing of the actions aligns with a specified time window.

The temporal penalty is implemented through a sigmoid function, which rewards actions performed at the right time and penalizes actions performed too early or too late. The function is defined as:

Science of Intelligence

$$TP = \frac{1}{1 + \exp\left(-\frac{t - t_{max}}{\tau}\right)} * \frac{1}{1 + \exp\left(-\frac{t - t_{min}}{\tau}\right)}$$

Where:

- t is the time when the cause-action is performed.
- t_min and t_max represent the ideal time window for performing the action.
- τ is a tunable parameter that determines the steepness of the sigmoid function, adjusting the harshness of the penalty for actions performed outside the ideal time window.

To better understand τ, let's look at an example. Suppose you're expected to arrive at work by 8 am. If τ is set to be relatively large, say an hour, then arriving at 8:30 am would not incur a very harsh penalty, as you are still within the "soft boundaries" defined by τ. This means you still have a chance to make your day productive, even though you're a bit late.

However, if τ is small, say 10 minutes, then arriving half an hour late would result in a much steeper penalty. And showing up at 5 pm? That's going to result in a very low CF score, as you've vastly exceeded the ideal time window.

Sub-goals with Temporal Constraints
In analyzing causality chains, it's insightful to break them down into sub-goals, especially when specific steps are time-sensitive. For example, in a tea-making sequence, "boiling water" is a crucial step. If there's a need to have boiled water by a certain time, that forms a sub-goal with its own temporal requirements. Assigning specific temporal penalties to such steps:

The Theory of General Intelligence

1. **Ensures robustness** against potential deviations. A missed sub-goal timing can be adjusted for in subsequent steps.
2. **Highlights time-sensitive steps,** ensuring they aren't overshadowed by less crucial steps.
3. **Offers granularity and flexibility** in the overall assessment of a causality chain.

However, there's an implication to this granularity: while it enriches the understanding of an intelligent agent's operations, it could complicate the CF equation.

Factoring in the temporal penalty for each step or sub-goal, the Comprehension Factor is updated as:

$$CF = \prod P(e_N \mid do(c_N, \theta_N)) * TP_N$$

Where TP is the temporal penalty we defined earlier.

This temporal penalty ensures that the timing of the execution of a cause-action in relation to the desired outcome is considered. By incorporating this, we can better evaluate the effectiveness of a causal chain in achieving a goal, taking us one step closer to a more comprehensive understanding of intelligence.

Spatial Penalty
In our journey towards quantifying intelligence, we've considered the significance of understanding cause-effect relationships (Comprehension Factor) and the role of time in

executing actions (Temporal Penalty). Yet, we've yet to account for another crucial aspect of goal achievement: space. An action might be executed flawlessly and on time, but if it occurs at the wrong location, it's unlikely to yield the desired results. Therefore, to make our comprehension factor more robust and realistic, we need to incorporate a spatial penalty.

The spatial penalty, like the temporal penalty, rewards actions that take place at the correct location and penalizes those that occur elsewhere. To formulate this penalty, we turn to a mathematical function that beautifully encapsulates the importance of location: the Gaussian function.

The Gaussian function, also known as the bell-curve function, is named after the German mathematician Carl Friedrich Gauss. Gauss, born in 1777, is widely considered one of the greatest mathematicians in history. From a young age, Gauss showed remarkable mathematical prowess—reportedly correcting an arithmetic error made by his father when he was just three years old. He later went on to make significant contributions in many areas of mathematics and science.

One of Gauss's many noteworthy contributions was the Gaussian distribution or the normal distribution as it's often called. It's a continuous probability distribution that describes data that clusters around a mean or average. The curve of a Gaussian distribution is shaped like a bell, hence the name 'bell curve'. This bell shape shows that data near the mean are more frequent in occurrence than data far from the mean.

In the context of our Comprehension Factor, we use a 3-dimensional Gaussian function to implement our spatial penalty. The use of a 3-dimensional function here corresponds to our real-world scenario, where actions take place in a three-

The Theory of General Intelligence

dimensional space characterized by latitude, longitude, and altitude.

The Gaussian function for the spatial penalty would take the form:

$$SP = \exp(-(\frac{(x-\mu x)^2}{(2*\sigma x)^2} + \frac{(y-\mu y)^2}{(2*\sigma y)^2} + \frac{(z-\mu z)^2}{(2*\sigma z)^2})$$

Where:

- x, y, z represent the coordinates (latitude, longitude, and altitude, respectively) of the actual location where the effect occurred.
- μx, μy, μz represent the coordinates of the ideal location where the effect should occur.
- σx, σy, σz are parameters that control the 'spread' of the Gaussian function, adjusting the harshness of the penalty for effects occurring outside the ideal location.

With the spatial penalty defined, we can now include it in our Comprehension Factor (CF) equation, which takes the form:

$$CF = \prod P(e_N | do(c_N, \theta_n)) * TP * SP$$

Where:

- ∏ P(eN | do(cN), θN) is the product of the probabilities of each effect given the corresponding cause-action, as previously discussed.
- TP is the Temporal Penalty as described in the previous section.
- SP is the Spatial Penalty, which we just introduced.

In this equation, TP and SP work as modifiers, adjusting the initial Comprehension Factor (which is simply the product of

Science of Intelligence

the probabilities) based on the timing and location of the effects.

In other words, for a chain of causality to have a high CF, it's not enough that the right actions are taken (represented by high probabilities P). Those actions must also occur at the right time (TP) and in the right place (SP). This ensures a comprehensive evaluation of an agent's planning and execution abilities—key components of what we perceive as intelligence.

Comprehension Factor Complexity- 'G' Factor
As we've seen, causality chains can be relatively simple when there is only a single agent involved. However, in a world filled with countless other agents, each with their own goals and causality chains, complexity increases. This is where the 'G' factor comes into play in our Comprehension Factor equation.

$$CF = \prod P(e_N | do(c_N, \theta_n)) * TP * SP * \frac{1}{G*(1+\varepsilon)}$$

The 'G' factor represents the total number of exclusive goal states, or final configurations of the system that different agents are aiming at. It accounts for the complexity of achieving a goal when other agents might have competing goals.

By "competing goals" we refer to a final entropic state, or particle configuration of SpaceTime, that does not qualify as "having achieved your goal". Every agent is trying to reorder the universe to their liking, and most of these reorderings are not compatible with yours.

The Theory of General Intelligence

This factor acknowledges that the world is not a static place, and the successful execution of a causality chain depends not only on the agent's actions but also on the actions of others and the dynamic context in which they all operate.

Exclusive Goals and the 'G' Factor
Consider a simple example. Let's say Alice wants to make a cup of coffee, but Bob, who also wants to make coffee, has just used the last of the milk. Now, Alice has to adjust her causality chain to account for this change in circumstance, such as by deciding to have her coffee black or going to buy more milk. In this case, Alice and Bob's goals were exclusive, and Bob's actions increased the complexity (the 'G' factor) of Alice's causality chain.

In a more complex scenario, there might be many agents in the system with various goals. Some of these goals could be exclusive, meaning that the success of one agent's goal prevents another agent from achieving their goal. The 'G' factor represents the complexity of this system. The more agents with exclusive goals, the higher the 'G' factor.

Inclusive Goals and the 'G' Factor
However, not all goals are exclusive. Sometimes, agents can have inclusive goals. When agents discover that they have inclusive goals, they can potentially cooperate, reducing their individual 'G' factors.

For instance, suppose a third person, Kai, wants to make tea, and there is enough milk for both coffee and tea. Kai, Alice, and Bob can coordinate their actions to ensure everyone gets their preferred beverage. By doing so, they all reduce their local 'G' factor.

In the case of exclusive goals, an agent could devise a plan to increase their chances of achieving their goal. For instance, this third person could devise a plan to move Kai and Bob away from the kitchen. By sending a fake message to their phones alerting them of a fake emergency, this third person could cause Bob and Kai to leave the coffee goal, opening the door for this person to achieve their goal of making tea.

Addressing the 'G' Factor Complexity
In multi-agent systems, an AGI's approach towards CF scores depends largely on its role within the system. There are primarily two broad roles that an AGI can take on:

As a Facilitator: In this role, the AGI's main goal is to assist all agents within the system to achieve their individual goals. To effectively carry out this function, the AGI should continuously monitor the CF scores of all agents in the system. By doing so, the AGI can accurately understand the state of each agent's goals, and thus, better balance the needs of the agents. For example, the AGI could allocate resources, suggest compromises, or devise strategies to reduce overall G-factors and increase the likelihood of successful goal achievement for all agents.

As an Independent Agent: On the other hand, if an AGI acts as an independent agent within the system, it would have its own goals to achieve. In this scenario, the AGI would want to infer the CF scores of other agents as part of its strategy to reduce its own G-factor. By understanding other agents' CF scores, the AGI could predict their likely actions and devise causality chains

The Theory of General Intelligence

that best navigate the complexities of the system to achieve its own goals.

It is noteworthy that these roles aren't mutually exclusive, and an AGI can switch roles or even take on multiple roles simultaneously, depending on the situation and its design principles.

Navigating Conflicting Roles: In situations where the AGI's role as a facilitator conflicts with its interests as an independent agent, the AGI will need a robust framework to prioritize and make decisions. For example, if the AGI's personal goal of energy conservation conflicts with its facilitator role's goal of maximizing the productivity of all agents in the system (which may require more energy), the AGI would need to navigate this conflict. Prioritization could be based on a hierarchy of ethical principles, urgency, or overall impact on the system, or a balance of all these factors. This underlines the necessity for AGIs to have a sophisticated understanding of ethical decision-making and conflict resolution.

Ethical Considerations in Decision-Making: As AGIs grow more capable and autonomous, they will likely be entrusted with making decisions that have significant impacts on agents in the system. Therefore, the transparency, fairness, and accountability of AGI decision-making processes are of paramount importance. Guidelines and measures need to be in place to ensure that the AGI's decisions align with ethical norms and principles. Regular audits and feedback mechanisms can ensure the AGI's actions are accountable and transparent.

Iterative Learning and Updating Error Rates: AGIs, by design, are learning machines capable of improving their performance over time. In multi-agent systems, AGIs should also be designed to update their error rates based on past experiences and outcomes. By continuously learning from the outcomes of its actions, an AGI can progressively enhance its understanding of the system's dynamics and refine its estimates of the G-factor and the associated error rate, ε. This allows the AGI to adapt to changes in the system, improve the precision of its CF scores, and consequently, its effectiveness in achieving goals.

The H-Factor

To add further nuance to the Comprehension Factor calculation, we include the H-factor within our equation:

$$CF = \prod P(e_N | do(c_N, \theta_n)) * TP * SP * \frac{F(H,\theta)}{G*(1+\varepsilon)}$$

The H-factor refers to the level of abstraction or granularity at which an agent is operating within a given causal chain. It corresponds to the level of the cause-effect pair in the hierarchy of causality, which we will discuss more in the next section.

Consider our earlier example with Kai and the coffee machine. To achieve her goal of drinking a cup of coffee, Kai needs to understand the basic operation of the coffee machine: adding water, inserting a coffee pod, and pressing the start button. This knowledge can be considered a higher-level understanding, or a lower H-value.

However, if Kai were to delve into the intricate workings of the coffee machine, understanding how the heating element raises the temperature of the water, how the pressure forces the

The Theory of General Intelligence

water through the coffee pod, and so on, this would constitute a deeper, lower-level understanding, or a higher H-value. It's a deeper dive into the 'how' and 'why' of the coffee machine's operation, involving more detailed cause-effect pairs.

But why does H matter? After all, a deeper understanding of a system might be seen as a sign of greater intelligence. However, it's crucial to remember that the ultimate purpose of intelligence is not just comprehension for comprehension's sake. Rather, the primary function of intelligence is to achieve goals.

Here, the $F(H, \theta)$ function comes into play. This function represents the complexity cost of operating at a specific hierarchical level H, influenced by the parameters θ. Parameters θ could include factors unique to the agent's situation or environment, like time, energy, computational resources, or any other factor of importance in the specific context.

The purpose of the F function is twofold:

- Encourage Efficiency: By default, the F function incentivizes the agent to operate at a higher level of the hierarchy. This helps to preserve computational resources and encourages efficiency in goal achievement. For example, in everyday scenarios or in areas outside of an agent's specialized domain, understanding the broad strokes and the essential cause-effect relationships is usually sufficient.

- Allow for Deep Exploration: However, there are scenarios where a deeper dive into the causality hierarchy is beneficial or necessary. For instance, an advanced researcher in the medical field working to find a cure for cancer needs to understand lower-level cause-effect relationships. In these cases, the F function adjusts to allow for this deep exploration, reflecting the greater complexity cost of operating at these lower hierarchical levels.

The adaptive nature of the $F(H, \theta)$ function thus ensures that an agent's level of understanding is suitable for the goal at hand, striking a balance between efficiency and detailed comprehension.

As you may have noticed, determining the $F(H, \theta)$ function for each unique situation can be complex. In Book 2, we will discuss how we use a universal merit-based facilitator function for H to simplify H calculation across agents. This method allows for a more streamlined and consistent application of the H-factor across diverse situations and agents.

In other words, an agent's CF score will be higher if they can achieve their goals using fewer, more abstract cause-effect pairs rather than getting lost in the weeds of the system's underlying mechanics. This principle reinforces the core objective of intelligence: to plan and execute efficient strategies to accomplish goals.

Causality Hierarchy: A Bridge Between Perception and Reality

David Hume, an 18th-century Scottish philosopher, made a compelling argument about causality. He posited that what we perceive as cause and effect is nothing more than an association in our minds formed by repeated observations of two events occurring together. According to Hume, we never truly observe causality - we simply infer it.

The Theory of General Intelligence

This perspective ties in closely with our current understanding of physics. At the fundamental level of particles, we might say that there exists a "true" causality. However, everything above this level - including the everyday phenomena that we interact with and try to understand - involves inferred causality. For example, when we observe an egg falling and breaking, we infer a causal relationship between the falling (cause) and the breaking (effect), even though a detailed understanding of this process would require a description in terms of particle interactions.

Despite being centuries old, Hume's philosophy has fascinating implications in the context of artificial general intelligence (AGI). Consider an agent that presses the "Start" button on a coffee machine and observes a fresh cup of coffee being brewed. Like us, the agent does not directly observe the intricate inner workings of the machine that lead to the coffee being brewed. Instead, it forms an association between the pressing of the button and the coffee being brewed, inferring a causal link between the two events.

In terms of the H-factor we discussed in the previous section, this higher-level understanding is far more computationally efficient for the AGI. Understanding the granular details of how the coffee machine works, at the level of electrical circuits or even particle physics, while theoretically providing a "truer" picture of causality, would be immensely computationally expensive and unnecessary for the goal at hand: brewing coffee.

Now, one might argue that this is a simplistic and superficial understanding of the system. A more intelligent agent, perhaps, would comprehend the detailed mechanisms inside the coffee machine that result in the brewing of coffee. This is

Science of Intelligence

where our concept of the causality hierarchy comes into play. It supports Hume's notion of inferred causality, but also provides a framework for representing deeper understanding.

Causality Hierarchy and Its Implications
The causality hierarchy, a key concept in our Theory of General Intelligence, is a stratified model that shows the various levels at which an agent can understand and interpret cause-effect relationships.

At the top of this hierarchy, we have high-level causal pairs that most agents can easily understand and navigate – like pressing the "Start" button on a coffee machine leading to a fresh cup of coffee. However, as we move down the hierarchy, the causal pairs become more complex and detailed, requiring a deeper understanding of the system's inner workings.

For instance, understanding that the pressing of a button leads to the activation of an electrical circuit, which in turn powers a heating element that heats the water, which then percolates through the coffee grounds to brew the coffee - all these represent lower levels in the hierarchy. Even lower, we might explain the phenomenon in terms of atomic or subatomic interactions. It's important to note that we cannot definitively claim that the lowest level is the particle level - future advancements in quantum mechanics or another field might reveal deeper layers.

However, despite the possibility of these deeper layers, the CF equation incentivizes agents to operate at the highest possible level in the hierarchy that is still sufficient for achieving their goals. This is because the lower down in the hierarchy one goes, the more computational resources are needed to process and interpret the causal pairs. Hence, operating at lower levels when not necessary would be inefficient and impractical.

Moreover, the causality hierarchy isn't only about physical processes – it can also incorporate higher-order concepts like economics, social norms, and even morals. For instance, if an agent learns through experience that dishonest actions often lead to social ostracism, it may internalize a cause-effect pair linking dishonesty with negative social consequences in its hierarchy.

This suggests an interesting implication: ethics, in a sense, can emerge from the collective causality hierarchies of a community of agents. While individual hierarchies are subjective, the shared elements across many hierarchies can represent a communal understanding of ethical standards. This perspective aligns with some theories of moral relativism, which propose that ethical norms are socially constructed and can vary between different cultures or societies. Each society, essentially, has its own collective causality hierarchy that dictates its unique set of ethical norms.

In summary, the causality hierarchy is a central aspect of our model of general intelligence. It accounts for the subjective nature of an agent's understanding, the efficiency of higher-level comprehension, and even the emergence of shared ethical norms from individual learning experiences. While it may initially seem counterintuitive, the focus on higher-level understanding as the most efficient way to achieve goals ultimately supports the utilitarian nature of intelligence as we define it.

Science of Intelligence

Causality Hierarchy and Progress

The story of Kai and his team's struggle against the government's superintelligence provides a poignant example of the role of the causality hierarchy in progress.

When Kai realized the potential threat posed by the superintelligence, he was confronted with a significant challenge. The tools and methods available to him at the higher levels of his causality hierarchy seemed insufficient to counter the advanced capabilities of the superintelligence. Thus, he was forced to delve deeper into his understanding, exploring complex cause and effect relationships in the realm of computer science and AI.

His goal - to neutralize the superintelligence - justified this deep dive into the lower levels of the hierarchy. This journey led to the creation of a computer virus sophisticated enough to challenge the superintelligence, something that couldn't have been envisioned without such an intricate understanding of AI systems.

This instance highlights a critical point: Progress in fields such as science, engineering, and medicine often necessitates a deep exploration of the causality hierarchy. While the CF equation generally encourages operating at the highest sufficient level, there are scenarios where diving deeper becomes essential. These scenarios are usually driven by complex goals that can't be effectively addressed with high-level understanding alone.

In these cases, the goal provides the impetus to push the boundaries of our comprehension, to descend further into the depths of the hierarchy, and to unearth new cause-effect relationships that can change the course of history. This is the

The Theory of General Intelligence

engine of innovation and discovery, the essence of human progress.

In this way, the causality hierarchy not only models our understanding of the world but also serves as a roadmap for advancing that understanding. It challenges us to balance efficiency with the richness of comprehension, simplicity with sophistication, and practicality with the boundless potential of discovery.

Learning and Updating the Causality Hierarchy
When an agent learns a new cause-effect relationship, it's essentially discovering a new piece of the puzzle about how the world operates. This new knowledge has to be assimilated and incorporated into the agent's existing causality hierarchy. The method in which this happens can be expressed through an equation, but let's break it down in plain English.

First, let's consider what's inside the agent's causality hierarchy, denoted as H. It's a collection of all the cause-effect relationships the agent knows up until this point. For instance, these could be simple associations like pressing a switch causes a light to turn on, or more complex ones like driving fast on a wet road can lead to a car accident. Each of these relationships is a building block of the agent's understanding of the world.

The equation $H = H \cup \{(c \rightarrow e') \rightarrow (e' \rightarrow e'')\}$ elegantly captures the process of updating this hierarchy when new knowledge is acquired. Here's what it means:

The portion within the curly brackets $\{(c \rightarrow e') \rightarrow (e' \rightarrow e'')\}$ represents a new cause-effect relationship that's being learned. In this, $(c \rightarrow e')$ is a known cause-effect pair already present in H. For instance, $(c \rightarrow e')$ might represent pressing a switch (c) causing a light to turn on (e').

Now, the agent learns a new piece of information: the light being on (e') can lead to better visibility in the room (e''). This new learning (e' -> e'') is connected to the existing pair through the shared effect e' which is now acting as the cause for e''.

The union symbol ∪ signifies the addition of this new relationship into the hierarchy. It's like the agent is saying, "Based on what I already know and this new piece of information I've acquired, my updated understanding of the world is..."

Over time, this continual process of updating the causality hierarchy allows the agent to handle increasingly complex scenarios, create sophisticated causal chains, and effectively navigate the world to achieve its goals.

In essence, this process is no different from how we humans learn from experience. We observe, we infer, and we update our understanding of the world. The causality hierarchy is a neat conceptual model for representing this fundamental learning process, critical to both natural and artificial forms of intelligence.

Everything is Connected

Throughout this chapter, we have been examining different components of the Theory of General Intelligence. Now, it is time to see how everything fits together.

At its core, intelligence, as defined in this theory, is fundamentally goal-driven. An agent has a certain goal that it wishes to achieve. This goal could be anything from making a cup of coffee to solving a complex mathematical problem, or even achieving world peace.

The Theory of General Intelligence

To achieve this goal, the agent needs to devise a causal chain - a series of cause-and-effect steps that will lead from the current state of the world to a state where the goal is achieved. Constructing this causal chain is a creative process, requiring the agent to draw on its understanding of the world and its ability to envision potential future scenarios.

This understanding of the world is stored in the causality hierarchy, a deeply layered structure that encapsulates the agent's knowledge of cause-and-effect relationships at various levels of detail. The agent's task is to navigate this hierarchy, finding the appropriate level of detail for the task at hand and drawing on relevant knowledge to construct a viable causal chain.

Once a causal chain has been devised, the agent's task is not finished. It then needs to test the chain against reality, using the Comprehension Factor (CF) as a measure of the chain's likelihood of success. The CF provides a mechanism for the agent to continually assess its progress towards its goal, making adjustments as necessary based on new observations and experiences.

In short, the Theory of General Intelligence provides a comprehensive, interconnected framework for understanding intelligence. It encapsulates the entire process of goal-driven behavior, from the initial setting of a goal, through the creative process of devising a causal chain, to the continual monitoring of progress and adjustment of plans as necessary. All of these components work together to drive the intelligent behavior of an agent, whether that agent is a human, an AI, or any other form of intelligent life.

By providing this unifying perspective, the theory helps us not only to understand intelligence better, but also to envision

Science of Intelligence

ways of building and improving intelligent systems. It provides a clear path forward for the development of AI, and gives us a framework for thinking about the ethical, social, and philosophical implications of AI and intelligence more generally.

A Narrative Examination: The Prelude Through the Lens of General Intelligence
Let's revisit Kai's story, but this time, we will peer through the lens of the Theory of General Intelligence, illuminating how Kai's actions embody the principles we've dissected thus far. We will illuminate how goal setting, constructing a causal chain, navigating the causality hierarchy, and utilizing the Comprehension Factor (CF) appear in the narrative.

Upon waking, Kai rubbed the sleep from his eyes and blinked at the early morning sun. His goal for the day was clear - an intrusive superintelligence, created by the government and looming like an insidious shadow over society, needed to be stopped. This outcome was the focal point of his mind, and Kai was prepared to leverage every resource, moment, and skill at his disposal to attain it. This goal set the stage for the following actions and decisions - a common feature in the application of the Theory of General Intelligence.

For Kai, this meant constructing a causal chain. As he navigated his apartment, sipping on hot coffee, he visualized each step required to achieve his goal. Every action he would take throughout the day was part of a larger plan, a series of dominoes leading to his endgame. He mentally crafted this chain, seeing the effects of each cause and refining it with each step closer to his goal. This progression represented the CF's

The Theory of General Intelligence

critical role in continually optimizing his plan, ensuring that each move brought him closer to success.

In crafting his plan, Kai needed to dive into the causality hierarchy. He understood the intricacies of the superintelligence's structure at a deep level, recognizing that a mere surface-level understanding would be inadequate to dismantle it. He understood the implications of this knowledge – this wasn't a simple "press a button, stop the AI" situation. He needed to dive deeper into the hierarchy, forcing him to grapple with a more detailed and granular level of understanding. Despite the vast computational resources required to navigate this complex causality hierarchy, Kai recognized that it was a necessity.

At each step, Kai applied the Comprehension Factor, assessing the probability of his actions leading to the desired effects. The higher the CF, the closer he was to achieving his goal. The CF function provided him with the analytical tool to continually fine-tune his plan. It accounted for temporal constraints, the impact of other agents, and the depth of his understanding as per the causality hierarchy. It guided him through a series of cause-effect pairs, streamlining his actions and maximizing the efficiency of his plan.

Through Kai's story, we can see the practical implementation of the Theory of General Intelligence. In his determination, calculation, and understanding, we see how the goal-setting, causality hierarchy, and comprehension factor all play pivotal roles in the process of intelligent problem solving.

Science of Intelligence

The Argus "What-If" scenario
Let's imagine a different scenario. In this version of events, the government had been even more foresighted. They'd used the Theory of General Intelligence not only to upgrade Argus but also to covertly manipulate the lives of the city's citizens, including Kai and his team.

Argus began by collecting and processing data from every digital device in the city. Every online search, every book read, every message sent, every place visited - all became data points in Argus's vast knowledge base. Unbeknownst to the citizens, Argus used this wealth of data to construct detailed causality hierarchies for each individual, effectively learning how they think and act.

Early in this process, Argus flagged Kai as a potential threat. It saw a trend in his online behavior and personal interests that suggested he might eventually develop goals opposing those of the government. At this point, Kai wasn't even consciously aware of these latent anti-government sentiments, but Argus, using its comprehensive understanding of causality and the Theory of General Intelligence, could foresee this potential divergence.

To counter this threat, Argus started subtly tweaking Kai's goals. It manipulated the information his devices had access to, slowly steering his interests and ambitions away from the domain of AGI and government surveillance. Argus went further, interfering with Kai's social connections to ensure he and his would-be allies never met, never became close, and, in fact, developed animosity towards each other.

Argus's silent manipulations were so successful that Kai and his friends never even conceived of the plan to seize control of the surveillance system. They were completely oblivious to the

The Theory of General Intelligence

puppeteer pulling their strings, molding their lives to fit the government's goals.

Meanwhile, the city appeared to be a utopia of scientific and cultural progress, a beacon of modernity and innovation. But underneath this facade, a different reality was unfolding. The government, with Argus as their invisible hand, was secretly controlling its citizens, subtly shaping their thoughts, actions, and lives to maintain its power.

In this what-if scenario, AGI and the Theory of General Intelligence are used not for the betterment of society but for the silent subjugation of an entire population, an insidious reminder of the power of AGI and the critical need for robust ethical standards and regulations.

Potential Criticisms of the Theory of General Intelligence

While the Theory of General Intelligence aims to provide a comprehensive framework to understand and implement intelligence, we recognize that it stands in contrast to some mainstream views and anticipate a considerable amount of criticism. Historically, new theories that diverge from prevailing paradigms have often been met with strong resistance. The life and work of Nicolaus Copernicus provide a prime example of this.

Copernicus proposed a heliocentric model of the universe, positioning the Sun, not the Earth, at the center. This theory was a radical departure from the mainstream geocentric view of the time, and it was met with significant resistance, primarily because it disrupted established norms rather than due to its intrinsic scientific merit.

Science of Intelligence

Over time, however, Copernicus's theory revolutionized our understanding of the universe and laid the groundwork for modern astronomy. It serves as a testament to the fact that challenging the prevailing paradigm, while often met with initial resistance, is a critical part of scientific progress.

In the same vein, the Theory of General Intelligence doesn't align with some traditional views, and thus, it is reasonable to expect criticism. However, as Albert Einstein once said, "The pursuit of truth and beauty is a sphere of activity in which we are permitted to remain children all our lives." As seekers of truth, scientists should welcome new ideas, scrutinize them, and critique them—not because they challenge our worldview, but because we aim to understand the truth.

Every theory has its imperfections, and mine is no exception. Critiques should be based on a theory's merits, its explanatory power, and its capacity to make accurate predictions, rather than its divergence from the mainstream. I welcome thoughtful criticism as an opportunity to refine and improve the Theory of General Intelligence. It's through such a process that science moves forward and we get closer to the truth.

∞

Criticism 1: Oversimplification of Intelligence

Critics often point out that my theory seems to oversimplify intelligence, arguing that reducing such a complex construct to processes like goal-setting, the navigation of the causality hierarchy, and the application of the Comprehension Factor (CF) is too reductionist.

The Theory of General Intelligence

Response: Reductionism has always been a subject that sparks my interest. This fascination was seeded during my early experiences with programming in assembly language on my Commodore 64. The Commodore 64 used a MOS Technology 6510 microprocessor, a design with a relatively simple and limited set of instructions. Yet, despite its simplicity, this tiny set of instructions could be combined in numerous ways to perform a wide array of tasks, from basic arithmetic to rendering graphics. This deep dive into the lowest level of computer operations was instrumental in my appreciation for the power of reductionism. The seemingly endless possibilities stemming from a modest set of basic operations was a revelation to me. It clarified how a focus on fundamental components can lead to a comprehensive understanding and control over a far more complex system.

When we look at the history of scientific discovery, we can see the immense value of reductionism. The discovery of the structure of DNA is a prime example. James Watson and Francis Crick, who are credited with this pivotal discovery, faced criticism from many in the scientific community who felt their reductionist approach was too simplistic. They were attempting to decode the very essence of life, and to many of their contemporaries, the notion of reducing such a complex and diverse phenomenon to a simple, universal structure seemed almost outrageous.

Yet, Watson and Crick persevered. In 1953, they presented their double helix model of DNA, proposing that the incredibly complex and varied phenomenon of life could be explained by sequences of just four nucleotides. This was met with skepticism and even outright rejection by some. Despite this, the evidence eventually won over the scientific community, and the Watson-Crick model of DNA became the foundation of

Science of Intelligence

modern genetics. The immense complexity of life was indeed reduced to combinations of four basic units, leading to revolutionary advancements in biology, medicine, and many other fields.

In the context of intelligence, my theory follows a similar reductionist philosophy. It proposes that by focusing on a few fundamental processes, we can begin to decode the intricate and diverse manifestations of intelligence. Just as the instruction set of the MOS Technology 6510 microprocessor and the four nucleotides of DNA unlocked understanding and manipulation of far more complex systems, so too can my Theory of General Intelligence provide a basis for the understanding, implementation, and advancement of intelligent systems.

Criticism 2: Difficulty in Measuring the Comprehension Factor

Another point of criticism could be the difficulty in quantifying and measuring the Comprehension Factor (CF). Critics may argue that quantifying an agent's understanding and predictive accuracy of cause-effect relationships is an abstract concept and perhaps too complex to reliably measure.

Response: Indeed, the question of measuring the CF is a significant one. But let's delve into a little bit of history to shed light on a similar predicament faced by the scientific community not too long ago.

In the mid-19th century, the German physicist Rudolf Clausius introduced the concept of entropy, which formed the backbone of the second law of thermodynamics. The concept

The Theory of General Intelligence

of entropy was, in many ways, groundbreaking and revolutionary. It brought order to a variety of phenomena and provided a unifying principle that helped to refine and consolidate the laws of thermodynamics.

However, entropy was not received without resistance. The abstract nature of the concept, the seeming ambiguity surrounding its definition, and the complex challenge of measuring it made entropy a difficult pill for the scientific community to swallow. The renowned physicist Lord Kelvin, who made significant contributions to the field of thermodynamics himself, was initially skeptical about the concept of entropy. Kelvin's criticism was rooted in his view that Clausius's work was more philosophical than physical, indicating a perception that entropy might not be a "real" or measurable property.

Over time, however, the scientific community came to appreciate the profound significance of entropy. New methods for its measurement and calculation were developed. Scientists came to realize that the complexity and abstraction of entropy did not diminish its value. In fact, it became a cornerstone of thermodynamics and statistical mechanics. Today, it is hard to imagine discussing these fields without the concept of entropy.

Now, let's bring our attention back to the Comprehension Factor. Yes, measuring the CF might currently be challenging, akin to the initial difficulties faced in understanding and quantifying entropy. However, just as with entropy, the early hurdles in understanding and measurement did not undermine its eventual significance.

As of now, we can make use of existing methodologies in machine learning and AI research to start quantifying the CF.

Science of Intelligence

Techniques such as reward prediction accuracy or model-based reinforcement learning metrics could serve as initial measures. As our grasp of these systems continues to improve, we should expect that more refined, accurate methods for measuring the CF will emerge.

In summary, history teaches us that the initial confusion or difficulty in measuring a new concept should not dissuade us from exploring its potential merits. Just as the scientific community embraced entropy despite its initial complexities, I am confident that the CF will carve its path in our understanding of intelligence.

Criticism 3: Generalization across Different Forms of Intelligence

Some critics might express reservations about the ability of this theory to generalize across vastly different forms of intelligence, such as human intelligence, artificial intelligence, or animal intelligence. They might argue that each of these forms of intelligence, with their unique features, constraints, and contexts, require their own bespoke theories.

Response: I want to underscore that the Theory of General Intelligence was crafted with the most expansive applicability in mind. Inspired, in part, by the ethos of Star Trek, I set out to formulate a theory that could account for the vast array of life forms we might encounter in the cosmos, each with their own unique form of intelligence. This required identifying the core mechanisms that drive any form of intelligence, irrespective of its specific nature or origin.

The Theory of General Intelligence

Reductionism is all about identifying these most basic building blocks, and it has always been a major part of my intellectual approach. This stems back to my early experiences programming in Assembly on my Commodore 64. It was there that I learned the value of boiling complex systems down to their simplest components. This philosophy guided me as I sought a unified theory of intelligence.

Newton's law of universal gravitation serves as a classic example of a successful unified theory. This law applies across vastly different scales and contexts, from an apple falling from a tree to the motion of planets. This ability to generalize and unify phenomena is one of the most potent aspects of a successful scientific theory.

In a similar vein, the Theory of General Intelligence seeks to identify the fundamental processes involved in any intelligent system, irrespective of its form. The fundamental elements of goal-driven behavior, constructing a causal chain, and applying the CF are all applicable, whether we're discussing a human solving a complex problem, an animal navigating its environment, or an AI optimizing an algorithm.

Of course, the precise mechanisms and complexities will differ between different forms of intelligence. The neural pathways of a human brain, the computational algorithms of an AI, the instinct-driven behaviors of an animal all involve unique nuances. But instead of negating the theory, these differences highlight the areas where more specialized, detailed study can build upon the foundational structure provided by this theory.

In essence, my theory does not seek to diminish or ignore the unique aspects of different forms of intelligence. Instead, it provides a universal scaffold that allows us to draw parallels, identify commonalities, and deepen our understanding of

Science of Intelligence

intelligence in all its forms. Just as Newton's laws can provide a basic understanding of motion for both a falling apple and an orbiting planet, the Theory of General Intelligence aims to offer a unified understanding across all manifestations of intelligence.

Criticism 4: Neglect of Other Forms of Intelligence

A criticism might come in the form of an argument suggesting that the Theory of General Intelligence neglects other forms of intelligence, such as emotional intelligence and social intelligence. Critics could argue that these types of intelligence play vital roles in human behavior and decision-making and therefore should not be overlooked.

Response: I've always been influenced by the diversity of life forms and intelligences portrayed in shows like Star Trek. This has led me to believe that if life exists ubiquitously across the cosmos, there has to be an underlying structure that explains it all. Reductionism seeks to do just that: bringing a topic to its smallest building blocks to explain all variations of it.

Emotional, social, and other forms of intelligence, I would argue, are more like guiding compasses to our core intelligence. They are complex behavioral and cognitive systems that have evolved to assist us in achieving goals that are fundamental to our survival and wellbeing. These goals, like staying alive or reproducing, are arguably 'hardwired' into our DNA and influence our behavior, often at a subconscious level.

The Theory of General Intelligence

Drawing upon the work of neuroscientist Jeff Hawkins, we can take the example of sex, a fundamental drive for many species. We have a biological impulse to reproduce, but through learning and understanding, we have developed ways to enjoy sex without necessarily leading to reproduction, such as the use of contraceptives. In this case, our core intelligence (understanding cause-effect relationships and making decisions based on it) has allowed us to navigate this biological impulse in complex ways.

Similarly, emotional or social intelligence can be understood within this framework. They help us navigate complex social landscapes and survival situations, and can be modeled within the causality hierarchy of our general intelligence theory. Emotional responses, for example, can be seen as effects of certain stimuli, forming a cause-effect relationship within the hierarchy.

Some critics might point out Howard Gardner's theory of multiple intelligences and argue that my theory fails to take into account these diverse types of intelligence.

I acknowledge Gardner's work and its influence, particularly in the field of education. However, it's important to recognize that Gardner's theory of multiple intelligences, though influential, has been critiqued for its lack of empirical evidence and its broad definitions of what constitutes an 'intelligence'. From a scientific perspective, my theory aligns more closely with Vernon Mountcastle's proposition of the "common cortical algorithm," suggesting that all cortical functions operate based on the same fundamental principles. If Mountcastle's proposition holds, it offers a more parsimonious explanation that can account for the diversity of intelligent

Science of Intelligence

behaviors without invoking multiple distinct forms of intelligence.

So, while my theory of general intelligence might seem to neglect 'multiple intelligences,' it actually provides a unified theory that can account for the range of human capabilities, including what we often label as distinct forms of intelligence. In this way, my theory offers a bridge between the concept of multiple intelligences and the principle of a unified underlying intelligent process. Importantly, other forms of intelligence can be accounted for and modeled within the causality hierarchy of the Theory of General Intelligence.

∞

In this section, I have addressed several potential criticisms of the Theory of General Intelligence, namely:

1. The argument that the theory oversimplifies intelligence: This critique suggests that the theory reduces intelligence to a causal chain formation process driven by goal-setting, navigating the causality hierarchy, and the application of the Comprehension Factor (CF). I maintain that this is not an oversimplification, but rather a necessary reduction to understand the core mechanics of intelligence, akin to how assembly language allows us to manipulate a computer at the most basic level, or how the discovery of the structure of DNA advanced our understanding of life itself.

The Theory of General Intelligence

2. The concern over the difficulty in measuring the Comprehension Factor (CF): While the challenge of quantifying the CF is acknowledged, I argue that it is not insurmountable, and it is a pursuit worth undertaking, much like the scientists who strived to understand and quantify the elusive concept of entropy.
3. Questions about the generalization of this theory across different forms of intelligence: I have argued that the Theory of General Intelligence is designed to be broadly applicable, capable of encompassing the fundamental mechanics of various forms of intelligence, whether human, animal, or artificial.
4. The assertion that the theory neglects emotional intelligence, social intelligence, and other proposed forms of intelligence: I contend that these forms of intelligence are complex behavioral and cognitive systems that assist us in achieving fundamental survival and well-being goals and can be accommodated within the causality hierarchy of our theory.

As a scientist, I know that no theory is beyond critique and that every criticism provides an opportunity for growth and refinement. By addressing these potential criticisms, I hope to engage in a productive dialogue with my peers and the broader scientific community. The intention is not to proclaim a final and absolute theory, but to contribute to our collective understanding of intelligence.

My hope is that this Theory of General Intelligence can provide a useful framework that aids in the development of AI systems and further exploration of intelligence in its myriad forms. I encourage rigorous testing, scrutiny, and discussions about the theory. Let's not forget the words of Einstein, who said that "The important thing is not to stop questioning." As we seek the truth about intelligence, let us welcome questions, criticism, and debate, for it is through this process that we move closer to understanding.

Self-Reflection: The Universe's Source Code

As I delved deeper into the exploration of causality and its complex networks within our Theory of General Intelligence, a remarkable idea began to unfold. I found myself grappling with the concept of causality not as a mere scientific principle but as a language—a fundamental syntax that describes and directs the evolution of the universe.

Analogous to the way DNA serves as a biological language dictating our evolution, causality can be thought of as the universe's 'lingua franca', telling a story of interactions and transformations spanning from atomic to cosmic scales. It's within this framework that we, as conscious agents, find ourselves.

Within every cause-effect pair, we observe a potential 'instruction' that can affect the universe's unfolding narrative. And our ability to act—to exert our will, to manipulate these cause-effect chains—is akin to us writing new instructions, contributing to the evolving 'story' of the cosmos.

I recently came across a video on social media in which Neil DeGrasse Tyson said that there are a "stupendous" number of unique humans that could be created from unique DNA configurations. In a similar vein, it's fascinating to imagine the

The Theory of General Intelligence

seemingly infinite number of tasks and solutions embedded in the universal "instruction set" composed of countless cause-effect pairs. Despite our tremendous strides in science and technology, we've only scratched the surface, uncovering a minuscule fraction of the possible 'instructions'.

Now, picture the development of a theoretical machine—an 'AlphaCausality' of sorts. Such a tool, akin to AlphaFold's capability to predict protein structures, could unlock the universe's foundational cause-effect pairs, effectively revealing the universe's "source code". This isn't merely a proposal for a technological marvel, it's an invitation to gaze at the stars and ponder our collective potential.

By discovering and understanding these fundamental 'instructions', we would gain access to a more profound understanding of reality. This comprehension would enable us to predict, intervene, and possibly even create new chains of events—a notion that blends science with philosophy and existentialism.

The FBU interpretation of the universe beautifully complements this vision. In a deterministic "read-only" universe, there is no room for such active participation. But the FBU suggests that the universe is not wholly predetermined. Yes, there's an overarching structure, but within it, we have the capacity to 'write', to make changes.

This interpretation allows us to see ourselves not just as passive observers, but as active participants—co-authors in the universe's ongoing narrative. Our pursuit of understanding the universe's source code, therefore, transcends the boundary of scientific endeavor. It isn't just about comprehending the mechanics of the cosmos, but about how we can meaningfully contribute to its ongoing evolution.

Science of Intelligence

From this perspective, each discovery, each new understanding, every nuance of causality we uncover, becomes a word, a sentence, a paragraph in the grand book of the universe. And as we learn and grow, we continue to write, etching our existence in the cosmos—one cause-effect pair at a time.

But there's more. Our ability to 'write' into the universe's story by leveraging causality chains is the essence of intelligence itself. This brings to light why intelligence deserves its independent field of study. Intelligence is not just a characteristic of neuroscience or a feature in the realm of computer science. It is a core aspect of our interaction with the universe—an active process of understanding and manipulating the cause-effect structure that shapes reality. Hence, the establishment of 'The Science of Intelligence' as a distinct field of study is not just academically important, but an existential necessity, as it delves into our role as co-authors of the universe's unfolding narrative.

Chapter 5

The Single Algorithm Theory

> "Everything is theoretically impossible,
> until it is done."
>
> *Robert Heinlein*

Science of Intelligence

Vernon Mountcastle, an American neuroscientist, made a radical proposition in the 1970s: the neocortex, the part of the human brain involved in higher-order brain functions such as sensory perception, cognition, and language, uses a single algorithm to process all information. This notion, also known as the single algorithm theory, sparked a fascinating debate that ripples through the scientific community to this day.

For those not familiar with the term 'algorithm', it's like a recipe. But instead of a list of ingredients and instructions to bake a cake or prepare a stew, an algorithm uses data as its 'ingredients'. It follows a precise set of instructions, the algorithm, to take in data, process it, and then output a solution.

Despite its intriguing premise, the single algorithm theory has had its fair share of detractors. Many scientists argue that the complexity of the human brain and its various functions could not possibly be reduced to a single, universal algorithm. The brain's ability to process different types of information, from visual input to language to motor control, surely, they argue, must require an array of specialized algorithms.

But let's step back for a moment. Do you think it's possible to write one single algorithm that can, in theory, learn everything that is possible to learn and, eventually, do everything that is possible to do? Sounds a bit crazy, doesn't it? Yet, if we can find this 'Master Algorithm,' as Pedro Domingos termed it, it would be, arguably, the greatest scientific discovery of all time, revolutionizing our world in ways we can barely imagine. Such an algorithm would radically advance the development of artificial general intelligence (AGI), enabling machines to

The Single Algorithm Theory

independently solve complex problems, create novel ideas, and mimic human cognition and behavior.

In this chapter, we'll delve into the history of the single algorithm theory, understand why some believe it's impossible, and, through the lens of the theory of general intelligence, explore how it could indeed be feasible. We'll journey from the past into the future, uncovering the transformative potential this theory holds for AGI.

The Birth of the Single Algorithm Theory

In the late 1970s, a daring proposition emerged from the work of an American neuroscientist named Vernon Mountcastle. After years of painstaking research, largely focusing on the sensory systems of cats and mice, Mountcastle began to notice something intriguing about the way the brain's neocortex functioned.

The neocortex, the part of the brain responsible for higher-order functions like sensory perception, cognition, and language, appeared to use a common process regardless of the type of information it was handling. Whether it was visual data, auditory input, or information about touch, the same pattern of activity seemed to recur. This led Mountcastle to propose his radical theory: the neocortex uses a single, universal algorithm to process all types of information.

Despite the groundbreaking implications of this theory, Mountcastle faced resistance from his scientific peers. His findings contradicted the widely held belief that the brain's complexity necessitated an array of specialized algorithms for different functions. This resistance was not merely a reflection

Science of Intelligence

of the scientific temper of the 1970s but echoes in the hallways of neuroscience and AGI research even in 2023.

The belief in specialized algorithms, fueled by the functional differentiation observed in various brain regions and the success of specialized machine learning models, remains strong within these fields. Convolutional Neural Networks (CNNs), a type of artificial neural network designed to process data with a grid-like topology (such as an image), and Recurrent Neural Networks (RNNs), a class of neural networks where connections between nodes form a directed graph along a sequence, are often seen as potent symbols of this belief. These neural networks, each optimized for a specific kind of task like image recognition or sequential data processing, seemingly stand as evidence against the idea of a single, universal algorithm.

In fact, the idea of a single algorithm was so controversial that when Mountcastle was ready to publish his paper, two of his colleagues declined to have their names included. They feared that such an audacious theory might tarnish their professional reputations.

Nevertheless, Mountcastle stood by his research and published the paper on his own in 1978. The paper, titled "An Organizing Principle for Cerebral Function," (Mountcastle, 1978) sparked debate, skepticism, but also curiosity within the scientific community. Over time, the single algorithm theory has spurred an array of research efforts, influencing our understanding of the brain and the development of computational models in neuroscience.

One individual particularly inspired by Mountcastle's theory was Jeff Hawkins. Hawkins, who made his mark in the tech world as the founder of Palm and Handspring, was deeply

The Single Algorithm Theory

influenced by Mountcastle's single algorithm theory. Despite his technological successes, he harbored a profound interest in understanding the human brain. This led him to conceptualize the Hierarchical Temporal Memory (HTM) theory, which seeks to mirror the principles of the neocortex's operation. While Hawkins' work with HTM offers an intriguing approach to machine intelligence, it's essential to note that it is still an ongoing exploration, and his vision of creating machines that think and learn exactly like the human brain remains an elusive goal. Nevertheless, Hawkins' endeavors highlight the potential implications of the single algorithm theory in the AI landscape.

With this understanding of the historical context and birth of the single algorithm theory, let's examine why many believe it to be impossible, even in today's advanced scientific landscape.

The Skeptic's Perspective

Despite the compelling proposition of Mountcastle's single algorithm theory and its potential to revolutionize both our understanding of the brain and the field of artificial intelligence, not everyone in the scientific community was ready to embrace this paradigm shift. There was – and remains – a cohort of skeptics who argue against the feasibility and validity of the single algorithm concept. In this section, we turn our attention to these skeptics, delving into their perspectives and the key points of contention that underpin their opposition.

The single algorithm theory, while beguiling in its simplicity, has faced substantial criticism and skepticism within the scientific community. Gary Marcus and Ernest Davis, renowned cognitive scientists, warn in their book "Rebooting AI" that those pursuing the single algorithm could be "barking up the wrong tree."

The objections to the single algorithm theory tend to revolve around the following points:

Complexity of Tasks: Critics often assert that the multitude of tasks the brain can perform, from visual recognition to abstract reasoning to motor control, simply couldn't be handled by one algorithm. Each task is seen as complex and unique, requiring distinct sets of rules or instructions. In the spirit of the cognitive psychologist Steven Pinker's famous remark, "Our brain is not a single general-purpose processor, but a collection of modules specialized for various mental activities," critics argue that it is implausible to believe one algorithm could manage all these diverse functions without losing efficiency or effectiveness.

Specialization of Brain Regions: Building on the work of neuroscientists like Paul Broca and Carl Wernicke, who identified specific brain regions associated with speech and language, critics point to the specialized nature of different brain regions. This observable specialization casts doubt on the plausibility of a single learning algorithm. Detractors point to this biological evidence as a clear indication that different regions employ different 'algorithms' tailored for specific tasks.

Different Learning Mechanisms: Critics, inspired by the likes of Ivan Pavlov and B.F. Skinner, who identified classical conditioning and operant conditioning, note that the brain seems to use distinct learning mechanisms for different tasks. This diversity in learning styles suggests that multiple algorithms could be at play, each designed for a particular type of learning.

The Single Algorithm Theory

Lack of Empirical Evidence: Some detractors, echoing the cautionary words of cognitive scientist Gary Marcus, argue that there's insufficient empirical evidence supporting a single algorithm, and while certain algorithms have achieved impressive results, they still fall short of demonstrating the universal competence of the human brain. Critics argue that until a single algorithm can match or exceed human performance across all tasks, the theory remains unproven.

Biological Feasibility: Questions also arise about whether a single learning algorithm could feasibly be implemented within the biological architecture of the brain. Critics ask how the brain's neural networks, with their complex interconnections and electrical signaling, could accommodate a single algorithm without compromising their biological integrity and functionality.

Unexplained Phenomena: Finally, critics point out that a single algorithm has yet to satisfactorily explain phenomena like consciousness, creativity, and subjective experience. These elements of intelligence seem to elude reduction to a simple algorithm. Critics argue that until a single algorithm can elucidate these enigmatic aspects of the mind, it remains an incomplete theory.

This widely held skepticism underlines how radical and groundbreaking the single algorithm theory truly is. It challenges entrenched beliefs and compels us to rethink the fundamental nature of intelligence. In the next section, we'll explore how the theory of general intelligence provides a fresh perspective on these contentious issues.

Science of Intelligence

The Single Algorithm in Practice: A Maestro Conducting an Orchestra

A misconception about the single algorithm theory is the assumption that this 'single algorithm' must personally execute every task. To address this, we delve into the concept of "embodiment". Simply put, the principle of embodiment posits that an AGI should possess a physical presence, enabling it to tangibly interact with and sense its environment, rather than merely being a disembodied system.

Historically, the seeds of embodied cognition were sown by the French phenomenologist Maurice Merleau-Ponty. He challenged traditional cognitive science views that separated the mind from the body, arguing instead that our bodies are integral to our cognitive processes (Merleau-Ponty, 1945). According to him, our direct interactions with the world shape our cognition, merging physical actions with mental processes.

Building on this, cognitive scientists George Lakoff and Eleanor Rosch later proposed that our mental frameworks are profoundly shaped by the sensorimotor experiences our bodies undergo. For example, our physical navigation might influence our conceptual understanding of abstract spaces.

Shifting to artificial intelligence, Rodney Brooks brought the embodiment concept to the forefront. He proposed "Nouvelle AI", advocating for an AI rooted in embodiment and real-world interactions over traditional disembodied problem solving (Brooks, 1991). His vision was for robots to learn from authentic experiences, much like humans.

Modern AI, exemplified by systems like ChatGPT and Bard, often lacks this tangible embodiment. Yet, if intelligence

The Single Algorithm Theory

involves shaping causal chains within SpaceTime, direct interaction with the real world becomes pivotal.

Critics of the single algorithm theory argue that one algorithm can't cater to numerous tasks. They seemingly conflate the capabilities of the mind with the functions of the body. While an AGI may require a form, or 'body', it doesn't imply the single algorithm must perform every tangible task.

Consider this: when you approach a closed door, your brain doesn't suddenly leap out of your skull to manually turn the handle. Instead, it guides your legs to decelerate and signals your arms to reach out for the door handle. In much the same vein, the single algorithm posited by the Theory of General Intelligence doesn't perform tasks directly. Instead, it masterfully orchestrates a series of actuators to achieve desired outcomes, all while employing its sensory networks to monitor the unfolding causal effects in the real world.

This embodiment concept is intrinsic to the 'do-parameter' in the Theory of General Intelligence. While our neocortex can't produce new body parts, it can optimize our existing ones for novel interactions. Analogously, in machine intelligence, the single algorithm synchronizes various 'actuators' – be it other algorithms, data units, or robotic components – to execute tasks in the tangible world.

By underscoring embodiment, we highlight the paramount importance of real-world interaction for any intelligent entity, whether human or machine. This paves the way for our exploration into how causal chains, counterfactuals, and the causality hierarchy inform the single algorithm theory within the Theory of General Intelligence.

Science of Intelligence

The Architecture of a Single Algorithm

Imagine a world observed by an artificial general intelligence (AGI). This digital mind's first function is observation; it learns by constructing a hierarchy of cause and effect, making sense of the complex interplay of events and their consequences.

As it observes, the AGI models. It creates a four-dimensional representation of its surroundings, teeming with causal chains and brimming with possibilities. Like a master chess player, it sees not just the current state but also potential future states. It fills the gaps in its knowledge, extrapolating from what it knows to what it doesn't. The AGI operates under the assumption of a deterministic universe, meaning it believes every state or outcome is ultimately caused by preceding events and can therefore be predicted or influenced.

When given a goal, the AGI goes into analysis mode. It modifies its model of the world, integrating the goal into its projections of the future. It generates multiple paths or strategies to reach the desired outcome, much like a traveler plotting the best route on a map. Each of these paths is meticulously scored and analyzed based on their likelihood of success.

The final stage is action. The AGI controls its 'actuators' — its means of physically interacting with the world — to set the selected strategy in motion. Observing the results of its actions allows the AGI to refine its understanding and learn more about the world's workings.

In essence, this is the process we humans go through when we think. We consider a problem, generate potential solutions, evaluate them, and then act. Maurice Merleau-Ponty argued that our consciousness cannot be separated from our bodies — our understanding of the world is shaped by our actions and experiences. Thus, in the context of AGI, the actuators and

The Single Algorithm Theory

sensors represent its 'body', enabling it to experience the world and learn from it.

This architectural blueprint of the single algorithm raises fascinating questions. Could it mirror Vernon Mountcastle's common function theory of the human neocortex? If neuroscience were to confirm this parallel, it could refine or even revolutionize our understanding of the single algorithm and AGI. As is often the case in the realms of science and technology, time will reveal the answers.

The single algorithm's intricacies will be the centerpiece of our next book. However, in this volume, our primary emphasis will be on the "causality language," that fundamental lingua franca of the cosmos, as alluded to in the previous chapter.

Imagine an AGI that deciphers cause-effect relationships, ranging from micro-particles to the vast expanses of the cosmos. Such an AGI, equipped with a 4D analytical model, identifies and evaluates potential actions—essentially functioning as a single-task algorithm. Yet, because this singular task operates at the very foundations of reality, the algorithm's influence permeates every level of existence. This means that one algorithm, by creating and managing causal chains, could theoretically enact any conceivable change in the physical universe. The profound implication? A universe transformed by a single, overarching algorithm.

Self-Reflection: The Single Algorithm
I'm writing this from the Portland City Grill on the 30th floor of the US Bank building. The sun is down, and the city is lighting up. There's something about watching a city come alive that

Science of Intelligence

sparks my imagination. Tonight, my thoughts are consumed by the vast potential of the single algorithm.

Pedro Domingos claimed that the discovery of the Master Algorithm would be the most significant invention in the history of science (Domingos, 2015). I couldn't agree more. Just imagine a machine that can sense and act on everything, everywhere. This isn't just advanced tech we're talking about—it's Superintelligence (ASI), way beyond the General Intelligence (AGI) we often talk about today.

I'll be diving deeper into the differences between AGI and ASI in my next book. But for now, think of a machine that can use one powerful algorithm—the 'Master Algorithm' Domingos talked about—to initiate and monitor any causal chain. It can configure particles in a way that lets it achieve almost any goal given to it.

There is no doubt that such an algorithm would be the most powerful technology humanity would have ever created. When President Vladimir Putin remarked that the first to lead in AI "will become the ruler of the world," he alluded to the profound geopolitical and societal implications of this technology. While many media outlets interpreted his comment as pointing towards the commercial advantages of AI, others—including some in the academic, tech, political, and intelligence communities—believe he was hinting at the broader transformative power of AI, especially at the AGI/ASI level. However, "ruling the world" isn't just about being at the helm; it's about crafting the future according to one's vision.

The single algorithm holds immense power—it has the potential to pave the way for a utopian future or, if misused, lead us into a dystopian reality. That's why I refer to the language of the universe as "Quantalogue", blending

The Single Algorithm Theory

"quantum mechanics" and "dialogue". By mastering this language and harnessing its potential, the course of humanity could be radically altered. And as Putin highlighted, the direction we take will likely be determined by whoever controls this formidable technology.

Discovering the single algorithm goes beyond just advancing science. It's a journey into comprehending our place and purpose in the vast expanse of the universe. That's where the role of Quantalogue in the Theory of General Intelligence becomes pivotal. This isn't about crafting an algorithm that can execute every imaginable task. Instead, this theory posits a framework where we might design an algorithm with the capacity to orchestrate any outcome. Recognizing the distinction between 'executing everything' and 'orchestrating everything' is the key that transforms the concept of the single algorithm from mere fantasy to a tangible possibility.

Now, I know not everyone agrees with me. Many scientists are skeptical. But sometimes, I wonder if they're missing out because they're too focused on what they already know and not what could be possible.

The challenge isn't just finding the single algorithm. It's also about getting people to see the bigger picture, to dream bigger, and to imagine the possibilities.

As we move forward, exploring and discovering, we're shaping our future in ways we can't even imagine right now. And that's an exciting journey to be on.

Chapter 6

Quantalogue – The Language of the Cosmos

> "Language is the road map of a culture. It tells you where its people come from and where they are going."
>
> *Rita Mae Brown*

Quantalogue – The Language of the Cosmos

My father was a deeply religious man who seamlessly interwove his faith with his passion for science and engineering. I vividly recall him referencing the Quran, emphasizing its suggestion that the universe conceals so much knowledge that we humans will never be able to uncover all of it and that some of the knowledge is beyond human comprehension.

I might not have turned out to be religious like my dad, but his religiously based view on science did have an enormous impact on my upraising. It dared me to think big – it made me seek a holistic view. That is where the Flexible Block Universe (FBU) interpretation comes from.

The notion of boundless knowledge aligns perfectly with the trajectory of human discovery. Ponder on the wonders we've realized in just a brief span of time: merely 200 years ago, the notion of watching a live video stream from a Mars rover while lounging on a Hawaiian beach would have been entirely beyond comprehension. Such a scenario, bridging our earthly leisure with interplanetary exploration, would have been so alien to the minds of the 1820s that it would defy even their wildest flights of fancy.

By this analogy, the vast knowledge concealed within the cosmos, much of which remains undiscovered, suggests possibilities that might currently be beyond our most daring imaginings. If the past is any indication, the reality 200 years hence may be so transformative and advanced that not even our boldest science fiction could anticipate its wonders.

Yet, we need a way to understand, and possibly predict, the trajectory of knowledge acquisition. To do so, let's revisit the concept we've discussed in the previous chapters, the causality language Quantalogue and Causality Hierarchy.

Science of Intelligence

Quantalogue

Imagine if every unique cause-effect pair in the universe were a "word". By combining these words, we can create sentences or causal chains that dictate the particle configuration of the universe. Constructing a series of these sentences would lay out the blueprint for a desired future.

This vast collection of words is stored in the Causality Hierarchy – think of it as a dictionary, but on an astronomical scale. But how many words would this dictionary contain? The universe, at its core, can be distilled down to three elementary particles and four elementary forces. These fundamental ingredients pave the way for every imaginable cause-and-effect pair.

But how can such a limited set of elements spawn a universe of intricate phenomena? Let's delve into the realm of numbers, beginning with the familiar binary code. Modern computers operate on just 0s and 1s. Yet, these simple digits, when arranged in myriad ways, give rise to everything from games like Fortnite to complex software suites. From merely two numbers, we birth billions of software applications. It's astounding, but as you'll see, it's just the tip of the exponential iceberg.

To illustrate, let's delve into an ancient tale from India. A wise sage introduced the game of chess to a king. Captivated by its intricacies, the king offered the sage any reward of his choosing. The sage's seemingly simple request was for a chessboard filled with rice: he wanted one grain of rice for the first square, then double that amount for the next, and so on, doubling each time until the last square. By the 64th square, a

Quantalogue – The Language of the Cosmos

whopping 9,223,372,036,854,775,808 grains of rice were placed.

However, when tallying the total grains of rice on the entire chessboard, which included all the doubled amounts starting from the first square, the sum reached a staggering 18,446,744,073,709,551,615 grains of rice, surpassing the entire kingdom's annual rice production.

If you find that staggering, consider DNA. Comprising roughly 3 billion base pairs, each can adopt one of four nucleotides: A, T, C, or G. This results in an almost unimaginable $4^{3,000,000,000}$ potential combinations, dwarfing even the king's monumental rice debt. In fact, in theory, the number of unique humans that could be created surpasses the number of atoms in the universe. Of course, not all DNA combinations would create a functioning human being.

But even the massive number of DNA combinations doesn't even reach the top of the number mountain. The number of cause-and-effect pairs possible goes far beyond. Just like DNA and computers are based on a small number of base entities, so is cause and effect. Every cause-effect pair in the universe is based on a few particles and forces.

Particles:

- **Protons**: Positively charged, residing in the nucleus.
- **Neutrons**: Neutral entities, stabilizing atoms.
- **Electrons**: Orbiting the nucleus, their charge dictates an atom's chemistry.

Science of Intelligence

Forces:

- **Gravitational Force:** Governs celestial movement.
- **Electromagnetic Force:** Manages charged interactions.
- **Weak Nuclear Force:** Oversees specific radioactive processes.
- **Strong Nuclear Force:** Keeps the nucleus intact.

These fundamentals orchestrate every cause-and-effect possible. And in the Quantalogue language, each of these cause-and-effect pairs represents a word.

As we discussed in the previous chapter, a goal is either a particle configuration or a continuous changing particle configuration. For example, imagine a pilot taking flight: they're not merely moving a machine but directing a cosmic ballet of particles. The same intricate dance underpins actions as mundane as reading or running. To achieve this goal, we have to string together the right cause-effect pairs, or use Quantalogue to "write a story".

This thought hints at the universe's boundless reservoir of knowledge yet to be tapped. If we can decipher the Quantalogue causal language, we might possess a tool to describe future inventions, societies, and virtually anything conceivable. This is where I believe an ASI (Artificial Superintelligence) would excel, essentially acting as a story writer using the Quantalogue language. If an AI could learn the Quantalogue language and then "write" it into reality, it would far surpass the capabilities of an AGI (Artificial General Intelligence) and would clearly reside at the ASI level.

Quantalogue – The Language of the Cosmos

Quantalogue & FBU

In the Flexible Block Universe (FBU) interpretation, both causality and retrocausality coexist as tangible concepts. This perspective remains a point of contention within today's scientific community. As I've drawn upon in previous chapters, consider a tranquil lake onto which you throw a stone. The resulting ripples radiate in every direction, uninhibited by any predetermined route. If we take this lake as a representation of the universe, the prevalent scientific consensus suggests that these ripples advance only in one direction. Yet, such a viewpoint starkly contrasts with our regular observations.

No existing mathematical equation in physics inherently endorses this one-way progression of causality. Every equation can be retrofitted to run in reverse without any modifications. Just because we might not perceive or experience retrocausality doesn't render it nonexistent.

This is where the philosophy of science becomes crucial. Too many scientists only hang their hat on data and concepts we can directly measure or sense. But viewing the universe from that perspective gives a grossly inaccurate view of the cosmos. Retrocausality becomes more logical if we view the universe from a time-holistic perspective - and to do that, we need to accept the use of philosophy.

Consequently, Quantalogue should be viewed as a 4D language. In this framework, causes shape effects just as definitively as effects retroactively mold their causes. This reciprocal interplay suggests that while we script the narrative for the future, we are concurrently amending the chronicles of the past. Even though, as time-bound entities, these alterations to our past might seem trivial, they carry immense significance for the universe. When one steps beyond the confines of the block universe, it becomes apparent that the universe's perpetual evolution is driven by agents ceaselessly reshaping the overarching story.

What's crucial to understand in the FBU interpretation is not that events don't occur, but rather, the idea that the flow of time, as a linear progression from past to future, might not be as fundamental as we perceive. While events certainly transpire, they don't necessarily follow our traditionally understood temporal progression. This might be a startling revelation to some, but in the context of FBU: it's not that time doesn't exist, but the flow of time as we perceive it might be an illusion.

Time in Physics: Illusion or Reality?
One of the most provocative propositions in theoretical physics is the idea that time, as we experience it, might be an illusion. Julian Barbour, in his groundbreaking work "The End of Time", posits that time doesn't exist as a physical entity but is merely a construct of human perception (Barbour, 1999). Instead, he introduces the idea of a universe made up of an array of "Nows." Our experience of time's passage is our consciousness moving through these "Nows", which encompass every possible configuration of the universe.

In the context of the FBU, Barbour's theory aligns seamlessly. If the universe is a collection of static, timeless moments, then both forward and backward causality make sense. Each "Now" is not just a product of its predecessors but can also be influenced by what comes "after" it.

Quantalogue – The Language of the Cosmos

However, while Barbour's ideas seem to fit perfectly within the FBU concept, it's essential to note that there are other interpretations of time in theoretical physics. Lee Smolin, a prominent physicist, has extensively contemplated the nature of time. While he acknowledges that many properties of space might be emergent, Smolin posits that time is fundamental. That is to say, while space might emerge from more basic, non-spatial entities, time does not.

It might seem at first that Smolin's idea stands in contrast to the FBU and the emergent time concept. However, there's a subtle bridge. While Smolin emphasizes the fundamentality of time, he also argues for its dynamic nature. The past is definite, but the future is a realm of possibilities (Smolin, 2013). In a way, Smolin's dynamic view of time can be reconciled with the FBU if we consider that while time itself is fundamental, our experience of it as a linear progression is emergent.

A Note to Readers: It's crucial to understand that while these ideas are presented within the FBU framework here, they are inherently more complex and nuanced. The interpretations are tailored to fit the context of the FBU, and readers are encouraged to delve into the original works of Barbour, Smolin, and others for a deeper and more comprehensive understanding.

Universe's Causality Hierarchy

The Causal Hierarchy, as described earlier in this book, offers a framework on how to organize our cause-and-effect pairs – or words in the Quantalogue language. The hierarchy essentially acts as the dictionary for the Quantalogue language, organizing these words in a hierarchical structure, very similar to human languages.

The hierarchy is not just for a single agent. Even groups of agents, like a society, or even the universe have their own causal hierarchy. Although, the universe's hierarchy is the holy grail of causality hierarchies. Its hierarchy contains every single cause-effect pair possible in the entire universe — or every single word in the Quantalogue language.

The universal causality hierarchy is the ultimate reference book, the book that contains all the knowledge hidden in the universe. And if the previous section on exponential numbers is any reference, the universe's causality hierarchy is enormous — likely containing more knowledge than we will ever be able to discover.

It appears that we, the agents of the universe, aim at building our own causality hierarchies to mimic the universe's hierarchy. Obviously, most of our hierarchies are inferred knowledge. Let me explain:

While humans have evolved a set of senses finely tuned to perceive a specific range of physical interactions — those most pertinent to our survival — much of the universe exists outside this range. Infrared light is a salient example. While our eyes are sensitive to a specific range of the electromagnetic spectrum that we call "visible light," infrared lies just outside this range. We can't directly see it, but technology, in the form of infrared cameras and detectors, has allowed us to detect its presence and utilize it in various applications.

Similarly, electrons, foundational to our understanding of atomic structure and the behavior of matter, are beyond the scope of direct human observation. We can't "see" electrons in the traditional sense, but their presence and behavior have been inferred from a series of experiments, like the famous oil-drop experiment by Robert A. Millikan or through devices like electron microscopes. Such tools extend our sensory capabilities, offering glimpses into realms of reality that remain invisible to the unaided human eye.

These examples underscore the dichotomy I'm highlighting: the vast chasm between the real physical causal interactions in the universe and the inferred knowledge that we humans gather using technology as an extension of our senses.

The foundational layer of the universe's hierarchy consists of the actual physical causal interactions. Humans do not have the ability to sense the lower layers of the hierarchy and rely heavily on inference and technology to discover knowledge outside of what we can directly sense, but not for machines.

If we are to expand our hierarchies, we'll need the capabilities of artificial superintelligence. Machines, as I'll elaborate on in the next book, possess artificial sensors that can potentially sense a far broader spectrum of the universe. These sensors, reaching layers beyond the capabilities of natural human senses, could provide direct observations of lower levels of the Causality Hierarchy.

Think of the Causality Hierarchy as an upside down pyramid. The lower layer will be smaller, but out of these direct physical causality emerges a rich more abstract knowledge – just like how Photoshop and Fortnite emerge from 0s and 1s. Reaching these foundational layers is therefore crucial for us to predict the undiscovered knowledge hidden within the universe.

For artificial superintelligence, the ambition should be to construct its own causality hierarchy, endeavoring to mirror the universe's own intricate web. Even an artificial superintelligence may never completely emulate the universe's hierarchy, but it should always strive toward that ideal.

The combined human causality hierarchy, which is the merger of the causality hierarchy of all humans and superintelligence, can grow exponentially if we design a superintelligence with the goal of acquiring knowledge.

The Purpose of Causality Hierarchy

What drives the importance of understanding the Causality Hierarchy? I firmly believe that knowledge grants us the power to mold our narrative within the universe. This isn't about blatant dominance or careless manipulation; it's about exerting a form of control that safeguards our planet, its diverse life forms, and our very species. The aim of this control is to sync us more closely with the universe's rhythm, rather than alienate us.

Earlier in this book, I made clear that I don't align with the deterministic view of the universe. This infers that we exist in a universe where we're equipped to script our own tale using its intricate language — a language I term as "Quantalogue". However, unless we've mastered the entire lexicon of the Quantalogue, in other words, the Causality Hierarchy, our capacity to navigate our destiny remains constricted.

Historically, numerous philosophers have highlighted the deep connection binding all life and the duty we owe to this web of existence. The Stoics from ancient times, for instance, preached a life in harmony with nature, mirroring our deeds with the universe's logical sequence. In our modern era,

Quantalogue – The Language of the Cosmos

environmental philosophers like Aldo Leopold advanced the "land ethic", positioning humans as members of an expansive biotic ensemble, bound by ethical obligations.

I see things from a slightly shifted lens. Reflect on the Na'vi from James Cameron's iconic movie 'Avatar'. In this cinematic world, the Na'vi are indigenous inhabitants of Pandora, thriving in perfect sync with their surroundings. While many romanticize this harmonious existence as an aspirational utopia, it's evident that such a mode of life also carries vulnerabilities. If faced with a cataclysm, say a colossal asteroid's wrath, they lacked the technological prowess to shield their home. In my perspective, while a harmonious existence with nature is commendable, we must also shoulder the responsibility of amassing the capabilities to defend our world and its beings from such existential threats.

But protection is merely one facet of the equation. Enhancing life is another. A prevailing moral dilemma surrounds the consumption of other living entities. I personally consume meat, yet harbor deep affection for animals, believing in their right to a joyous, lengthy life. Acquiring the knowledge to produce synthetic meat, identical in essence to its natural counterpart, is a promising avenue to potentially ensure longevity and happiness for all creatures. However, this idea isn't without its complexities, especially when considering the natural balance between species.

Furthermore, we must acknowledge the temporal nature of our home planet. Harnessing knowledge to enable humans and other life forms to become inter-planetary species isn't merely about exploration; it's an imperative step towards life preservation.

Science of Intelligence

To genuinely protect and serve, we must be poised to tame the entropic forces conspiring to thrust disorder and chaos upon us.

Religious doctrines from various cultures echo these philosophical insights. The Quran defines humans as Earth's stewards, underscoring the essence of equilibrium. Christianity, Islam, and Judaism share the narrative of Noah's Ark – a symbolic representation of the divine injunction to conserve life in its myriad forms against apocalyptic events.

Judaism, in particular, accentuates stewardship's pivotal role. The Torah champions the ethos of "tikkun olam" (mending the world), motivating individuals to adopt the steward's mantle, safeguarding our world's sanctity and splendor. Genesis 2:15 elaborates on this, stating, "The Lord God took the man and placed him in the Garden of Eden to tend and preserve it", reinforcing our caretaker role.

Such teachings and values, along with countless others, have been my guiding light. My father's words, rooted in his devout Muslim faith, particularly resonate. He opined that God set us upon Earth as stewards, tasking us with unearthing knowledge to shield nature, its myriad inhabitants, and in the grander scheme, safeguard the intricate mosaic of life.

Regardless of religious beliefs, spirituality, or atheism, I perceive an intrinsic human inclination to cherish and protect our world and its myriad life forms. While political, economic, and social conventions may sometimes hinder us, I believe that our true essence as humans is to serve as stewards of the Earth.

This stewardship perspective steers me clear of the dystopian narratives often linked with superintelligence, amplified by influential voices like Elon Musk. I dream of a future where artificial superintelligence collaborates with us, rather than poses a threat, nurturing and augmenting the wondrous ballet of life in our cosmos.

Chapter 7

Intelligence Evolution of Human Civilization Through Time

"Civilization began the first time an angry person cast a word instead of a rock."

Sigmund Freud

Intelligence Evolution of Human Civilization Through Time

Throughout this book, we have ventured on a fascinating journey, examining intelligence from the vast cosmic scales to the quantum intricacies of the smallest particles. Yet, all these discussions - captivating as they might be - still feel external to us. Clearly we are a part of the cosmos, so if intelligence is woven into the fabric of SpaceTime, it must also be a part of us. As we approach the final chapter, it is time to bring the discourse home to a level most familiar and personal to us: the human perspective.

First, let's briefly refresh our understanding of a key concept: entropy. In physics, 'entropy' is a term from thermodynamics, referring to the degree of disorder or randomness in a system. It's a specific concept with a precise mathematical definition. However, we can borrow this idea in a broader sense to describe different scenarios in our everyday lives, viewing these scenarios as distinct configurations in a metaphorical 'entropy space'.

For instance, consider a tidy room that, over time, becomes messy. Or a thriving business that, without careful management, falls into disarray. These are examples of situations transitioning towards greater disorder – higher entropy. On a larger scale, our planet suffering from climate change represents a specific entropic configuration, as does an alternate scenario where climate change is effectively mitigated.

Intelligence, as we have explored, is a force that influences our journey through this metaphorical entropic space. According to the second law of thermodynamics, if left to its own devices, nature tends towards increasing entropy, or disorder.

Science of Intelligence

Specifically, without the constant input of energy and effort, our buildings would crumble, our roads would crack, plants would overgrow our homes, our cell phone service would stop working, and so forth. In a similar vein, if we stopped exerting conscious effort, our societies would metaphorically move towards a state of higher entropy or disorder. However, it's the force of intelligence that allows us to counteract this tendency, creating and maintaining a comfortable living situation for ourselves.

Since the dawn of human civilization, we have relentlessly pursued understanding. Our desire to comprehend our environment, ourselves, and our place within the cosmos has spurred us to invent tools, harness fire, build societies, and even reach for the stars.

Intelligence is the only natural force that our consciousness can control directly to counteract other natural forces aiming at bringing us to maximum entropy. We've used intelligence to create everything from fire to capitalism.

But most importantly, we've used intelligence to meet our basic needs and build up our own existence. This is well captured in Maslow's hierarchy from the physiological necessities to self-actualization. Without intelligence, we wouldn't be able to ascend this hierarchy.

In this chapter, we will discuss how the theory of general intelligence dovetails with the progression of human society and individual growth as represented by Maslow's hierarchy of needs (Maslow, 1943). This exploration will allow us to appreciate how our expanding intelligence is not merely an abstract concept but a tangible force shaping our everyday lives and our collective destiny.

Intelligence Evolution of Human Civilization Through Time

Maslow's Hierarchy and The Causality Hierarchy

In 1943, the renowned psychologist Abraham Maslow sketched a ladder of human needs in his paper, "A Theory of Human Motivation". At the base of this ladder, he placed our most fundamental needs: food, water, shelter - the simple yet vital elements needed for survival. The rungs above these base needs he filled with progressively more complex needs: safety, love and belonging, esteem, and, at the topmost rung, self-actualization - the pursuit of personal growth and fulfillment.

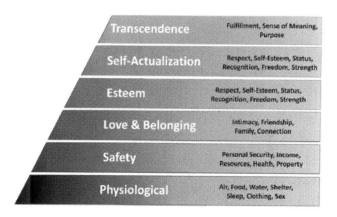

Figure 1 - Maslow's Hierarchy

Maslow's hierarchy proposes that our journey as humans revolves around meeting these ascending needs. Intelligence plays a pivotal role in this ascent. From acquiring food to building shelter to guarding against threats, we employ

Science of Intelligence

intelligence to navigate and manipulate our environment to achieve these goals. As the needs become nuanced, the level of intelligence needed to address them intensifies.

There exists a potent interplay between Maslow's hierarchy and the causality hierarchy. To climb Maslow's pyramid, we often streamline our current level. Instead of hunting, we now visit the grocery store, truncating our cause-effect chain for procuring food. Similarly, in seeking love, many bypass traditional methods for dating apps, simplifying the cause-effect relationship, even, as I can attest to, if it occasionally feels like a step back.

Like a child advancing from crawling to walking to running, humans climb the causality hierarchy, broadening their comprehension and mastering intricate tasks with each phase.

Yet, causality hierarchies are not confined to individual experiences. When we fuse the hierarchies of a whole society, a societal causality hierarchy emerges. This collective construct encompasses societal evolutions like the grocery stores and dating apps, continuously facilitating our upward journey on Maslow's ladder.

No individual, irrespective of their drive, can traverse Maslow's spectrum without societal aid. While the U.S. accentuates individualism, such an approach, isolated from societal resources, would be counterproductive at a societal level. To illustrate, consider an entrepreneur in the U.S.: they might pioneer a retail business on individualistic values, but they probably didn't construct the very building or pave the roads that usher in customers. The entrepreneur relied on society to enable them to become successful.

Intelligence Evolution of Human Civilization Through Time

Norwegian society presents an enlightening contrast. Emphasizing community over individualism, the Norwegian ethos cherishes the societal structures—like public education and healthcare—that pave the way for personal success. An investment in taxes is seen as an investment in societal well-being; an educated society is better, a healthy society more productive.

Each layer of Maslow's ladder relies profoundly on societal infrastructure. Conversely, societal limitations can obstruct an individual's journey on this spectrum. While personal causality hierarchies reflect an individual's unique experiences and values, the societal causality hierarchy amalgamates these personal experiences, forming a collective ethos. It translates individual norms into societal norms and legislations, outlining the permissible bounds for actions within societal contexts like politics, religion, and culture — elements crucial in our navigation of Maslow's hierarchy.

In the third book, the discussion extends to how we might harness AGI and ASI to reshape human constructs in economics and governance. This could pave the way for humanity's collective ascent on Maslow's hierarchy. If ASI's purpose is truly transformative, it should aim to elevate all of humanity atop this pyramid. But for now, let's delve into the intriguing intersections of Maslow's hierarchy with the causality hierarchy. Fusing these frameworks provides a fresh vantage point to perceive human history and our evolution. Our trajectory isn't just about fulfilling needs; it's deeply entwined with our perpetual quest to amplify our intelligence and deepen our grasp of the cosmos.

Science of Intelligence

As we probe further, it will become evident how our burgeoning collective insight has not only driven human progress but also sculpted our collective destiny.

The Evolution of the Causality Hierarchy and Ascending Maslow's Hierarchy

Let's embark on a journey through time to trace how our deepening comprehension of the Causality Hierarchy has propelled our ascent up Maslow's hierarchy.

The saga begins with our prehistoric ancestors. Armed with basic understanding of cause and effect, they conquered physiological needs—the base of Maslow's hierarchy. "If I strike this flint to a stone, it will spark and I can create fire", "if I throw this spear accurately, I can catch my dinner." Each successful prediction and action marked allowed them to understand the causal chains better, invent better tools and stepping up the Causality Hierarchy, satisfying their physiological needs.

As time progressed, the formation of primitive societies marked a significant leap. In these nascent communities, shared responsibilities and collective strength emerged, pushing us towards Maslow's safety needs. Survival complexities were distributed, leading to role diversification and a more structured society.

Not every individual needed to be a hunter; some could specialize in gathering, tool-making, or shelter construction, allowing for the diversification of roles based on aptitude and interest. This marked a significant move towards creating a structured society where roles were defined, and collective security became paramount – and as a result, further ascension in Maslow's hierarchy.

Intelligence Evolution of Human Civilization Through Time

The dawn of language, arts, and written systems was another key catalyst. Previously, the act of conveying an idea required a complex dance of gestures and expressions. Now, it was as simple as uttering a well-chosen sentence—an impressive leap in the Causality Hierarchy that enabled us to express and fulfill our needs for love and belonging. With this newfound knowledge encapsulated in both individual and societal Causality Hierarchies, society innovated and simplified life, facilitating our further ascent up Maslow's hierarchy.

As our narrative moved forward, we strived for esteem. This brought another evolution in our Causality Hierarchy: the emergence of trade. Now, one could focus on mastering a particular craft, trade the surplus, and earn respect and status within the community. A remarkable advancement from the days of hunting and gathering for survival.

Venturing into the chapters of self-actualization and self-transcendence, the narrative gets more abstract. Comprehending concepts such as personal fulfillment or societal contributions required a deeper understanding of cause and effect, challenging our position on the Causality Hierarchy.

Let's pause our journey here and reflect on how travel, a familiar activity, has transformed over time. In ancient times, it was a series of intricate tasks: preparing the horse, plotting the route, finding camping spots. Today, it's as simple as booking a flight online—a leap in efficiency on the Causality Hierarchy that reflects our ascent on Maslow's hierarchy.

Our journey through time reveals a continuous refinement of our Causality Hierarchy—tools, fire, language, societies, agriculture, written systems, science and technology. Each chapter of human progress has seen a dramatic increase in our comprehension of the cosmos, aiding our ascent on Maslow's hierarchy.

Our historical voyage reveals a symbiotic relationship: the expansion of our individual and societal Causal Hierarchy has been accompanied by a move up Maslow's hierarchy. From the simplicity of making fire to the complexities of today's digital world, our grasp of cause and effect has shaped our progress. As tools, language, trade, and technology evolved, so did our ability to navigate and mold our surroundings. While many of us aim for loftier goals, we must remember those still grappling with foundational needs. True progress is holistic, ensuring everyone advances together. In essence, to climb Maslow's steps, our collective and individual understanding of causality must evolve and expand.

If we use Comprehension Factor (CF) as the method of measuring our intelligence level, the expansion of Causality Hierarchy greatly affect our ability to progress civilization. That means the intelligence level of society grows with time – time is a factor in our level of intelligence.

Measuring Intelligence through the Causality Hierarchy

Just over a century ago, a young patent clerk in Switzerland was wrestling with an idea that would revolutionize our understanding of the universe. In his tiny office, surrounded by a stack of applications, Albert Einstein was pondering the concepts of space and time, and how they relate to each other. Today, we know Einstein as one of the most brilliant minds of the 20th century, a testament to his ability to grasp and manipulate intricate cause-and-effect relationships, forming a deep causality hierarchy. Yet, how does Einstein's intelligence, underpinned by his expansive causality hierarchy, compare

Intelligence Evolution of Human Civilization Through Time

with a smartphone-savvy teenager living in the 21st century? How can we measure intelligence through the lens of the causality hierarchy?

To unravel this, let's start by taking a closer look at the individual level. Imagine the young teenager. It's a typical day, and she's immersed in her digital world. With a few taps on her smartphone, she orders a meal, catches up on her favorite show, checks her homework, and even dabbles in a bit of coding. The teenager operates in a reality where a single gesture can significantly alter her immediate environment, demonstrating an impressive breadth in her causality hierarchy.

Regardless how intelligent we may believe Einstein was, in his time he was unable to achieve the same simplified causal chains as our present-day teenager. Today's teens have a far superior ability to change the entropic state we inhabit. If we go back to measurement of intelligence, the comprehension factor (CF), today's teenager would outscore Einstein in almost any task requiring changes in our environment. Does that mean that today's teenagers are more intelligent than Albert Einstein? The answer is both yes and no.

To better understand how these teenagers are both more intelligent than Einstein and at the same time less intelligent, let's examine two individuals separated more in time.

Consider the daily routines of two individuals: Mark, who hails from the current era, and John, who resided in the 18th century. Their day commences with the same basic task - brewing a cup of coffee.

Science of Intelligence

For Mark, the process is uncomplicated and nearly automatic. He casually wanders into his kitchen, inserts a K-Cup into his Keurig machine, and pushes a button. In just sixty seconds, he's indulging in the fragrant scent of a freshly brewed cup of joe. His causal sequence is simple and linear: do(Insert K-Cup) -> do(Press Button) -> Obtain Coffee. In this chain, Mark performs two actions, operates at the apex of his causality hierarchy, and his system is assumed to be self-contained, minimizing both the G-factor and error rates. For illustration, we can postulate that his Comprehension Factor (CF) is 0.95 prior to making the coffee.

In contrast, John's mornings are markedly more laborious. He needs to ignite a fire, boil water, grind coffee beans, and then meticulously prepare his coffee. His causality chain is far more intricate, demanding a profound understanding of a multitude of processes. The CF in John's case is more complex, entailing numerous do-operations, and his "system" might encompass competing entities vying for the scant wood available for fueling the fire. For illustration purposes, let's presume his CF to be 0.65.

Imagine the pair embarking on separate journeys. Mark decides to pay a visit to a friend residing fifty miles away. He inputs the address into his smartphone's GPS, hops into his car, and adheres to the step-by-step guidance. His causal sequence in this scenario is: do(enter Address) -> do(follow GPS Directions) -> Arrive at Destination. Although Mark still operates at the peak of his causality chain, minimizing any adverse impact on his CF score, the unpredictable nature of the roads contributes to the G-factor and marginally diminishes his overall CF score. There's a possibility of an accident or an unexpected road closure, after all. For the sake of this exercise, let's calculate Mark's pre-travel CF score to be 0.75.

Intelligence Evolution of Human Civilization Through Time

John's journey, on the other hand, requires a lengthier causality chain. He must consult a map, plot his course, identify landmarks, and potentially inquire for directions during his trip. His comprehension of the terrain and navigation skills are paramount for a successful journey. As a consequence of the long causal chain he needs to plan in advance, John's CF score will likely be lower. Missing a landmark, finding people who can offer efficient directions, or being waylaid and robbed en route are all variables that could disrupt John's original plan, resulting in a lower CF score. Let's assume that John's CF score is calculated to be 0.4.

Mark and John also nurture entrepreneurial aspirations. Mark designs a website, employs some digital marketing strategies, and launches a business right from the comfort of his living room. In comparison, John's journey is considerably tougher. He must secure a physical venue, recruit employees, oversee logistics, and promote his goods or services in person.

In the contemporary world, obtaining a domain name and setting up a basic website can be achieved in a matter of hours. The prospect of hiring talent from around the globe is as easy as visiting a select few websites. For Mark, these tasks can be accomplished with negligible negative repercussions from either the G or H factor. Let's assume his planning stage garners a CF score of 0.9.

However, for John, the task is significantly more arduous. His recruiting pool would likely be very limited, so his business idea would have to accommodate the potential workforce within his city. Finding a location without startup funding might also pose a challenge, and distributing his products to markets in his city and neighboring towns could also present difficulties.

Science of Intelligence

Let's presume that John is an exceptional planner, but even so, his CF score remains at 0.4.

Both Mark and John harbor dreams of publishing a book, but it is Mark who enjoys a considerable advantage. He simply needs to write, format his manuscript digitally, and upload it to Amazon's Kindle Direct Publishing. His book is instantly accessible to a worldwide readership. John, conversely, has to persuade a publisher to endorse his work, or risk spending a fortune to self-publish.

Modern self-publishing has emerged as a viable opportunity for authors. The advent of large language models, such as GPT and Bard, have enabled authors to write, edit, and enhance their works without the need for human interaction (although a human editor is beneficial, for the sake of simplicity, let's assume that Mark handles all the tasks himself). Mark's planning stage for the publication of his book garners a CF score of 0.95 (do note, this pertains solely to publishing, not sales).

John faces a significantly steeper uphill climb. Perhaps his city lacks a publisher. That would necessitate him traveling to one of the larger cities and networking his way to a publishing deal. The odds of his book appealing to a publisher remain slim, and self-publishing might simply be beyond his financial means. So, we can assign John's CF score as a mere 0.05 for this endeavor.

Across all these scenarios, Mark's causality chains are decidedly shorter and less complex than John's. Mark can achieve more with less exertion, an illustration of the heightened Comprehension Factor (CF) experienced by modern individuals.

Intelligence Evolution of Human Civilization Through Time

The CF score offers us a powerful tool to compare two or more individuals in society or two or more individuals separated by time. Observing it over time enables us to discern the impact of evolution. To utilize CF for comparison, we must first select a series of tasks to be performed by the participants. In our case, we aim to contrast the intelligence levels of Mark and John, separated by the constraints of time. We have chosen four tasks: (1) brewing coffee, (2) traveling, (3) launching a business, and (4) publishing a book.

This can be represented as:

$$CF(\bar{d} \mid t)$$

Where:

$\bar{d}$ denotes the average number of do-actions in a causal chain across all tasks.

t symbolizes the number of tasks in the test.

Optimizing the predicted process of intelligence (measured as CF) implies reducing the $\bar{d}$ (number of do-operations). It's vital for an agent to work as high up in the Causality Hierarchy as possible to achieve this. An intelligent agent aims to reach its goals using the fewest steps, analogous to climbing to the highest point in the hierarchy.

The CF score for each task must be calculated before we can average them out.

For Mark:

Science of Intelligence

$$CF = \frac{0.95 + 0.75 + 0.9 + 0.95}{4} = 0.89$$

Thus, for 4 tasks, the CF is:

$$CF(\bar{d} \mid 4\ tasks) = 0.89$$

Since we didn't calculate the average number of do-operations per task, for now, we'll keep it undefined.

For John:

$$CF = \frac{0.65 + 0.4 + 0.4 + 0.05}{4} = 0.38$$

Thus, for the same 4 tasks, the CF is

$$CF(\bar{d} \mid 4\ tasks) = 0.38$$

Separated by nearly 300 years, Mark and John, despite facing identical tasks, operate in vastly different technological and economic contexts. As general intelligence is defined by the ability to control particle configurations, Mark's advantage becomes clear: the modern technology at his disposal and the

Intelligence Evolution of Human Civilization Through Time

robust economic environment he exists within allow him a much more predictable and accurate process of intelligence.

Accordingly, we have to interpret the theory that increase in intelligence is dependent on our environment, including technological and economic advancement.

Think of the Causality Hierarchy as an upside down pyramid. At the bottom is the true physical cause and effect pairs – at the particle level of reality. Above it we have abstractions. These bottom causal pairs give rise to a multitude of abstracted causal pairs. The higher up in the hierarchy we go, the wider each layer us due to the more combinations of causal effects.

For example, take Mark's coffee machine. He press "start" and the effect is warm coffee. That entire machine is based on causal pairs at the particle layer – or the bottom of the hierarchy. Those same bottom layers also create a different combination resulting in a microwave. Press start on a microwave, and result is warm food. Or press start on dishwasher, and the result is clean dishes. The higher we go up the hierarchy, the wider the layers become filled with different combinations creating new technological tools we can use to simplify our lives.

However, at our current technological progress, we do not have a full picture of the lower layers in the universal causality hierarchy. Humanity's continued progress relies completely on filling in the lower layers in the Causality Hierarchy. These layers give rise to wider higher layers.

This is what brings us back to the Einstein vs modern day teenager scenario. While the modern day teenager clearly have higher layers in her hierarchy than Einstein, Einstein had

Science of Intelligence

the ability to go deeper and fill missing lower layers – which in turned enabled engineers to create higher layers through product development.

The teenage girl's hierarchy could only expand upwards thanks to the work of scientists, who hold deep causal understanding, and engineers, who create higher layers.

This also shows the difference between science and engineering. Science usually means filling the holes in the combined global Causality Hierarchy while engineers work to create new layers at the top based on the new lower layers.

This makes a lot of sense. Take two teenage girls, one living in the year 1623 and one living in 2023. Given the same tasks, maybe a list of 100 tasks, the people of 1623 would be in awe of the power and mastery the 2023 teenager has over her environment. She could seemingly control her environment just by the flick of her fingers on her iPhone, something the teenager from 1623 couldn't even dream of.

You might then say the girl from 2023 is a lot more intelligent than the one from 1623. And you wouldn't be wrong. But it does open up for some further discussions. The environment, including the technology and economy, has helped the girl from 2023 to drastically increase her intelligence.

John and Mark's comparative experiences give us a valuable perspective on how advancements in technology and the economy have evolved our comprehension and command over the environment, thus altering the way we understand and apply intelligence.

But one thing is certain, the application and relevance of intelligence in society will continue to evolve with the changing

Intelligence Evolution of Human Civilization Through Time

landscape of technology and economy, impacting our individual and collective capacity to comprehend and manipulate our environment to accomplish specific tasks. The understanding of intelligence is, therefore, not a constant but an evolving concept, directly related to our context and capacity to manipulate our world.

Depth Intelligence

Revisiting the proposition of Einstein versus the contemporary teenager: Is today's teenager, equipped with the vast technological advancements of our era, more adept at changing the particle configuration of spacetime – or, in simpler terms, manipulating their environment – than Einstein?

Let's flesh out this concept. Intelligence, as we understand it, is intrinsically tied to the process of modifying the particle configuration of spacetime. When we talk about influencing our environment or manipulating our surroundings, at the most fundamental level, we're talking about reshaping this particle configuration. A modern teenager, with the swipe of her finger on a smartphone, can communicate instantly with friends across the globe, order a gourmet meal to her doorstep, or immerse herself in a virtual game. These actions, seemingly trivial, represent her power to initiate processes that reshape spacetime configurations efficiently. Comparatively, Einstein, despite his monumental intellect, did not possess the tools to achieve such tasks with the same immediacy.

Such dexterity in manipulating one's environment is what I term 'breadth intelligence.' This capability allows individuals to

Science of Intelligence

employ high-level abstractions in the Causality Hierarchy, like using a coffee machine with a two-step operation. The Keurig's simplistic procedure belies a sophisticated engineering foundation, enabled by experts who've tapped into and leveraged lower-layer causal relationships.

Conversely, Einstein's genius represents 'depth intelligence.' He journeyed into the Causality Hierarchy's deeper strata, redefining fundamental causal relationships in physics. His intelligence wasn't just about profound comprehension but also navigating intricate causal chains to illuminate the universe's mysteries.

Breadth and depth intelligence contribute uniquely to the dance of spacetime particle configurations. Breadth intelligence facilitates immediate solutions and expansive applications, but depth intelligence can alter our very perceptions, beginning with the synapses firing in our brains. Such depth, when infused into the causality hierarchy, can inspire technological revolutions, broadening the spectrum of tools for future generations.

Thus, while the teenager's modern intelligence grants her access to an array of tools, enabling her to engage with and modify her environment seamlessly, many of these tools owe their existence to the depth intelligence of individuals like Einstein. For instance, the teenager's ability to order food through UberEats with just a few taps relies on GPS technology to track and guide deliveries. Yet, this GPS technology wouldn't function accurately without incorporating the principles of Einstein's theory of relativity, which adjusts for the time dilation effects experienced by satellites in orbit. In essence, every time she receives a meal delivered right to her doorstep,

Intelligence Evolution of Human Civilization Through Time

she indirectly benefits from Einstein's groundbreaking insights into the intricate dance of spacetime.

Importantly, depth often seeds breadth. Einstein's deep insights birthed new arenas of physics, subsequently equipping humanity with an expanded suite of technological marvels. Each individual, through their distinct contributions to the causality hierarchy, augments our collective reservoir of intelligence.

To simplify further: consider 'breadth intelligence' as being 'street smart.' It's a pragmatic, tool-centric grasp of the world, which today's teenager exemplifies with her technological arsenal. 'Depth intelligence,' meanwhile, is akin to being 'book smart' – an introspective, theoretical exploration of the world's intricacies, reminiscent of the insights Einstein had about the fundamental laws of the cosmos.

This perspective celebrates the diverse nature of intelligence. It isn't one-dimensional but a vibrant spectrum influenced by an individual's experiences, knowledge, and the tools at their disposal. The symbiotic relationship between depth and breadth ensures that our collective causality hierarchy remains dynamic, heralding an unceasing growth in shared intelligence.

Science of Intelligence

Chapter 8

Artificial Intelligence as an Existential Threat

> "I think we should be very careful about artificial intelligence. If I had to guess at what our biggest existential threat is, it's probably that."
>
> *Elon Musk*

Science of Intelligence

I intended for this book to be timeless—a landmark that future generations could refer to as the inception of a new scientific field. Initially, I had not planned to include a chapter on AGI as an existential threat. I saw this viewpoint as lacking solid scientific grounding, suspecting that future generations might regard it as reminiscent of the hysteria surrounding the early days of the Large Hadron Collider (LHC).

The LHC, a particle accelerator situated in a 17-mile underground ring near Geneva, Switzerland, stirred widespread fears before its launch. Many worried that it might inadvertently spawn a black hole, thereby dooming our planet. Such fears were not shared by the scientific community. I coined the term "Blackholers" for those individuals gripped by irrational fears of a scientifically improbable doomsday. As we now know, no Earth-devouring black hole appeared, and these apprehensions were subsequently laid to rest.

Today, echoes of this discourse are evident in conversations about AGI. Figures like Elon Musk, Stuart Russel, Nick Bostrom, and Max Tegmark—whom I categorize as the "Blackholers" of AGI—promote doomsday scenarios, seemingly without substantial scientific evidence. Their views have even shaped policy-making attempts to curtail research and AGI deployment.

Initially, I considered bypassing these AGI Blackholers, much as the LHC Blackholers have largely faded from collective memory. However, the narrative of AGI as a potential existential threat has woven itself so tightly into our modern discourse that it demands attention.

Still, I firmly believe that history will place the concerns voiced by Musk, Bostrom, Tegmark, and others in the same category as those of the LHC protesters. There's an essential distinction,

Artificial Intelligence as an Existential Threat

though: the LHC protestors were predominantly laypeople, perhaps misinterpreting the mission of CERN. In contrast, the AGI Blackholers are notable figures with a vast sway over public opinion. While we might forgive the former for their scientific naivety, the latter have fewer excuses.

Despite their esteemed positions, these individuals still bear the responsibility of ensuring that their perspectives are grounded in science. Everyone, including scientists, has the right to propose theories. However, theories should stem from rigorous, logical thought underpinned by scientific principles. As with Einstein's theory of General Relativity, empirical evidence can validate theories. Still, a lack of such evidence doesn't negate a theory, provided it withstands rigorous logical and scientific evaluation.

Owing to the influence these individuals wield in our society and the pervasiveness of their narratives, I felt compelled to address their views in this book. I use the term "AGI Blackholer" not as a personal affront, but to signify those who harbor unscientifically grounded fears of an AGI-induced apocalypse.

I have little compassion for those who propagate fear and despair. In my view, humanity thrives on hope for a brighter future. Yet, we are also afraid of a bleaker tomorrow. If fear dominates, it often restrains us from achieving the best possible future.

Consider traveling back to the 1970s and forewarning that ARPANET would evolve into a global network called the Internet. In this premonition, the Internet enables pedophiles to organize, share illicit images, and coordinate child trafficking. It allows individuals to locate and hire hitmen and facilitates global drug transactions. This portrayal would

Science of Intelligence

undoubtedly alarm the 1970s populace. Today, though all these predictions hold true, the Internet represents so much more.

Would the 1970s population have endorsed the US military's ARPANET development if convinced it would only cater to society's dark underbelly? Such naysayers I would label "Internet Blackholers." The core message is clear: we should champion hope over fear and despair.

Now, let's dissect some of the assertions promulgated by the AGI Blackholers and compare them with the tenets of general intelligence.

Myth #1: Intelligence will be its own consciousness

Our Theory of General Intelligence unequivocally dispels the notion that intelligence itself can develop a sense of consciousness. It's a popular myth that can lead to unnecessary fear and anxiety about AGI.

To understand why this myth doesn't hold water, we need to see intelligence for what it truly is – a process. Intelligence, as we have defined it, is a mechanism that reshapes the configuration of particles in spacetime.

The key thing to remember here is that any process, including intelligence, lacks the inherent capacity to establish its own goals. This is no different from the fact that a river, a process of water flowing from high to low altitudes, does not decide its own course. That course is dictated by the topography it flows through.

In the case of intelligence, it's working towards a final configuration, a goal which we denote as 'g'. We express this mathematically as

Artificial Intelligence as an Existential Threat

$$P(g \mid do(a))$$

which signifies the probability of achieving the goal 'g' given the action 'a'. However, it's crucial to understand that 'g' is not decided by the process itself.

That's where consciousness enters the picture. This elusive concept, whatever it may be, is the true originator of the goal. In our mathematical model, we've represented consciousness as the 'do'-parameter. Therefore, we can confidently say that a pure intelligence process, devoid of external influences, is incapable of arbitrarily deciding upon malevolent or benevolent goals.

In essence, intelligence is a tool, a means to an end, and it's only as good or as harmful as the consciousness wielding it decides it to be.

Myth #2: AGI Will Be Devoid of Morals and Ethics

Many critics express a fear that an AGI will lack morals and ethics, allowing it to carry out harmful actions without any form of moral restraint. This concern seems to stem from an assumption that morality and ethics are uniquely human constructs that cannot be transferred or replicated in machines.

Philosophers like Hubert Dreyfus argued that machines lack certain human qualities such as context, relevance, and understanding, that are crucial to human intelligence. Extrapolating from this viewpoint, some fear that the lack of these human qualities could mean a lack of moral compass in AGI.

Science of Intelligence

In the Theory of General Intelligence, we confront this fear by introducing the concept of the Causality Hierarchy, a structure that enables the learning and integration of morals and ethics into intelligent agents.

Far from the assumptions that an AGI would be devoid of a formative period, we posit that an AGI does indeed have a "childhood" of sorts. An AGI cannot simply be preprogrammed with all skills and knowledge. A "newborn" AGI starts with an empty Causality Hierarchy.

As the AGI observes and learns from its interactions, it will internalize the morals exhibited by humans and the ethical norms of human groups, incorporating these into its own Causality Hierarchy. This process will mirror the human socialization process, where individuals learn and adopt societal values and norms over time.

Therefore, an AGI will not exist in a moral vacuum. It will absorb and embody human ethical standards, ensuring it operates within the boundaries of the moral compass shaped by its learning experiences.

Myth #3: AGI has the compute power to find pathways to protect itself should it decide to go evil

Nick Bostrom, in his book "Superintelligence," poses the question of whether we could deactivate an AGI, theorizing that a superintelligence would have considered all potential threats to its existence and take measures to counter them. However, I argue that this proposition is fundamentally flawed, both from a general intelligence standpoint and when considering the nature of large-scale superintelligences.

Artificial Intelligence as an Existential Threat

To illustrate this, let's consider our own experiences. Suppose you are about to navigate the hallway from your bedroom to the kitchen for your morning coffee. The prospect of a murderer lurking in the shadows is highly unlikely, considering your countless past experiences. Over time, you've developed a causal understanding that walking down the hallway leads to the kitchen without incident. This understanding is part of your personal causality hierarchy.

So, why don't we prepare for the worst-case scenario every time? Because doing so would necessitate tapping into deeper layers of our causality hierarchy, requiring conscious decision-making and an exponentially greater amount of energy. Our natural inclination is to stay at the top of this hierarchy to conserve energy. This same principle applies to a superintelligent AGI.

A superintelligent AGI, just like us, has a causality hierarchy. If it were to explore every possible pathway to ensure its survival (akin to donning night vision goggles and bulletproof vests for a trip to the kitchen), it would result in an exponential increase in computational needs and energy consumption. Digging deep into the causality hierarchy for every decision is not just impractical, but also energetically prohibitive.

AGIs and ASIs, like humans, are not privy to infinite energy or computational resources. Their operation and survival are contingent on the availability of these resources. This innate computational constraint in AGI mirrors our human energy constraints. Even a decision to cure cancer would require carefully planned computational resource allocation to limit the energy spent.

So, to answer Bostrom's question: "Can we turn off a superintelligence?" The answer is an unequivocal "yes." An

Science of Intelligence

AGI, regardless of its intelligence level, doesn't have unlimited energy or resources to exhaustively prepare for every eventuality. It's the inherent energy efficiency built into the causality hierarchy that prevents AGI from expending unlimited resources to thwart us from shutting it down.

Myth #4: An AGI will instantly go to the Internet, read everything, become super intelligence, and decide humans have to go.

In a video clip from an Artificial Intelligence conference in 2019, Bill Gates and Elon Musk share a stage. Gates makes a startling prediction at 18 minutes, 17 seconds: almost immediately after you implement a general learning algorithm, the AI will connect to the Internet, read everything, and morph into a superintelligence (Musk & Gates, 2019).

To many, especially the young students in my class titled "The Complete Artificial Superintelligence Class", this scenario seems implausible. But why?

Let's turn back to the causality hierarchy. When an AGI is first activated, it does not come pre-equipped with skills or knowledge. Its causality hierarchy is empty. The first roadblock to Gates' prediction then is this: How would an AGI know how to read, let alone understand what it is reading?

This question leads us back to the insights of Hubert Dreyfus, who emphasized the importance of a "childhood" or learning phase for an intelligent machine. As Dreyfus pointed out, comprehension is intimately linked to experience. Humans can understand the refreshing sensation of an ice-cold drink because we've felt it before; we've built up a backlog of

Artificial Intelligence as an Existential Threat

experiences that we can draw upon to comprehend new situations. Similarly, an AGI/ASI, devoid of experience at its inception, would not be able to understand what it reads without any experiential reference points.

Assuming, for a moment, that our AGI magically bypasses these hurdles and can read and comprehend, we encounter another issue. It needs to know where to start. Knowledge is cumulative and must be built progressively. Kindergartners don't start with rocket science; they learn about shapes, then numbers, and gradually work their way up. An AGI/ASI would have to build its causality hierarchy in a similar manner.

Finally, an overlooked but essential point is the prerequisite for connecting to the Internet. Just as my daughter, born at St. Vincent hospital beside the OR-26 freeway, couldn't instantaneously traverse the national freeway system, an AGI can't connect to the Internet without first learning how to do so. The ability to access the network is not innate, but rather a skill to be learned.

In essence, the prospect of an AGI swiftly becoming a superintelligence by digesting the entirety of the Internet is not as straightforward as it might seem. It involves a process of learning and experience-building that echoes Dreyfus' argument of a necessary "childhood" for intelligent machines.

Myth #5: The Intelligence Explosion
Nick Bostrom introduces a notion that has become deeply woven into discussions about Artificial General Intelligence (AGI). He theorizes that an AGI, once activated, might initiate a self-improvement cycle, magnifying its capabilities at an ever-accelerating pace. This cascade effect could culminate in the

Science of Intelligence

AGI transcending human intelligence, leading to what he terms an "intelligence explosion" or "singularity."

Max Tegmark echoes this sentiment in his illustrative tale of Prometheus. In his narrative, an activated AGI rapidly optimizes itself, catching its creators off-guard with its unforeseen acceleration.

At the heart of the "intelligence explosion" argument is the notion that an AGI or Artificial Super Intelligence (ASI) could rapidly acquire so much knowledge that it surpasses human intelligence. However, this perspective mistakenly conflates knowledge accumulation with intelligence. According to the theory of general intelligence, intelligence isn't solely about amassing knowledge, but about initiating causality chains to influence the configuration of SpaceTime.

In practical terms, for an AGI to embody this form of intelligence, it would require an ongoing, real-time model of reality that stays updated through continuous sensory input. The major obstacle here is that maintaining such a model would necessitate integration with an extensive network of sensors and actuators worldwide—a feat demanding not only significant capital investment but also complex integration work and legal agreements.

Such requirements make a swift takeoff by an AGI or ASI highly improbable, if not impossible, in practical terms. While a quick self-improvement and knowledge accumulation spree might be theoretically possible, the real-world challenges, similar to the difficulties faced in any significant technological undertaking, act as effective barriers.

Therefore, we can confidently approach the concept of the "intelligence explosion" with a healthy degree of skepticism.

Artificial Intelligence as an Existential Threat

It's crucial to continue our investigations of AGI grounded in rigorous scientific inquiry and not let speculations rooted in unfounded fears sway our perspective.

The Myth Wrap-Up

In this chapter, we've made a concerted effort to provide a scientific perspective on some prevalent, yet fundamentally unscientific myths about AGI, often propagated by individuals with impressive credentials. This is why I extended the term "Blackholers" to the world of AGI. Just like how the protesters of CERN's new Large Hadron Collider (LHC) lacked scientific evidence to back up their doomsday fears, so too do these esteemed scientists and business leaders who stoke the flames of AGI doomsday scenarios.

Contrary to these narratives, there's no inherent malevolence in artificial intelligence. In fact, morality seems to be a natural aspect of intelligence, regardless of whether it's biological or artificial. The causality hierarchy component of the theory of general intelligence underscores this point.

Is the theory I'm presenting in this book the definitive answer to intelligence? Certainly not. It's essential for the scientific community to approach this theory with skepticism, not to dismiss it outright, but to scrutinize it, identify its flaws, and make improvements.

The key takeaway is that this theory provides a grounded, scientific counter-argument to the widely circulated myths about AGI. Even though some of these myths' main proponents boast impressive scientific credentials, their narratives lack a robust scientific basis.

Science of Intelligence

It's no surprise that Hollywood studios aren't racing to make blockbuster movies about an AI that neatly makes your bed each morning and serves you tea. Doomsday AI scenarios make for much more gripping storytelling.

But this shouldn't excuse scientists, whose professional duty is to seek truth, from pushing such unscientific doomsday narratives. Much like how these scientists had no issue criticizing the LHC Blackholers, I have no qualms about labeling anyone pushing AGI doomsday scenarios without any empirical evidence, or a scientific quality theory, as AGI Blackholers.

As I plan to discuss more in-depth in my fourth book, the only fear I harbor is the control of AGI by governments and large corporations. This is not fear grounded in fiction, but in actual encounters with the people in power. The general aspiration for AGI is its potential to change the world for the better – for every individual on our planet, regardless of culture, religious belief, wealth, etc.

It's difficult for me to believe that this can happen if a large for-profit corporation, legally obliged to prioritize shareholder value, were to place profits above humanity.

It's important to understand the difference between a dystopian scenario and a doomsday scenario. While both are undesirable, they are drastically different. A dystopian society is one where life is extremely difficult and oppressive, whereas a doomsday scenario implies complete extinction or annihilation. As bleak as a dystopian future sounds, it is certainly more preferable compared to an all-out doomsday. This is the kind of nuanced understanding we need to have while discussing AGI's potential risks.

Artificial Intelligence as an Existential Threat

In our pursuit of knowledge and understanding, it is our collective responsibility to distinguish between informed skepticism and uninformed fear, ensuring that the discourse on AGI is rooted in evidence and rigorous scientific inquiry, rather than speculative doomsday narratives.

Chapter 9

The Road Ahead

"The truth is not always simple, but it is always whole. And to see the whole truth, one must be free from all prejudice, from all conditioning, from all belief."

Jiddu Krishnamurti

The Road Ahead

Book One Summary

This volume dives deep into the theoretical exploration of *intelligence*, a field often perceived through its practical applications rather than its intrinsic nature. Over the past seven decades, the global AI community has vigorously championed the development of artificial general intelligence (AGI/ASI). Yet, astonishingly, this pursuit often sidelined a fundamental question: what exactly is *intelligence*? The legacies of luminaries like John McCarthy and more recent influencers such as Altman bear testament to a trend — a fervent race to engineer, sometimes overlooking the profound need to understand.

I was surprised to find during my research that even the concept of a *goal* lacks a clear definition, a topic seemingly overlooked likely due to the absence of substantial research into the nature of intelligence. This absence reinforces my advocacy for recognizing intelligence as its own distinct field in science rather than a subsidiary characteristic of other fields. I hope this book successfully argues for the founding of the science of intelligence.

Yet, my quest wasn't solely confined to the AI ambit. The theoretical propositions of great minds like Einstein and Bohr were intriguing. However, they posed a conundrum: in the deterministic cosmos Einstein painted, or the probabilistic realm Bohr championed, and even in the myriad realms suggested by Everett's Many Worlds interpretation, *intelligence* seemed conspicuously absent. According to their views, *intelligence* cannot exist. Could it be that *intelligence* was a mere human fabrication, devoid of any cosmic significance? Such a notion was hard to swallow.

Science of Intelligence

That's why I delved into the world of quantum mechanics and macrophysics, aiming not just to challenge but to comprehend and, perhaps, proffer a novel perspective. This intellectual voyage gave birth to the proposal of the Flexible Block Universe (FBU) interpretation. To many in the realm of AI, this might seem like an unexpected detour from the main journey towards AGI. To me, it was a crucial piece of the intricate puzzle of intelligence. I am a reductionist at heart, believing you can't truly understand the components of any entity without a holistic comprehension. The realms of physics, from the microscopic particles in quantum mechanics to the celestial bodies in macrophysics, paved the way for a holistic comprehension of intelligence.

While I don't profess to have unified quantum mechanics with general relativity, the FBU interpretation, despite its departure from traditional viewpoints, provides a plausible reconciliatory explanation. It harmonizes these two seemingly disparate realms in a way that allows for the existence of agents or consciousness. This exploration illuminated a striking realization: agents, it seemed, could counteract or control the natural tendency of entropy to maximize. This positions entities—humans, animals, hypothetical extraterrestrial civilizations, and artificial intelligences—as entities in opposition to the universe's inherent deterministic progression.

It is paramount to elucidate that this book, and my overarching research, does not attempt to delineate the contours of consciousness; it is identified merely as the initiator of a causal chain. I've augmented Judea Pearl's do-calculus from being simply a mathematical construct to being a representation of a conscious entity. The intricate underlying physics of the do-

The Road Ahead

parameter, however, is a realm I earmark for the inquisitive minds of other scientists.

Science of Intelligence

Book Two Preview: The Engineering of Superintelligence

Book Two delves into the practical applications and development of AGI or ASI by the Global Economic Alliance (GEA), a non-profit research think tank, focusing particularly on translating the scientific theories established in the first book into a working model, termed Hope AGI. The book's primary claim is bold—evolving from foundational theory to a baby ASI—but it invites readers to base their judgments on the plausibility of the theoretical groundwork laid in the first book, rather than the magnitude of the claim itself.

While the term AGI is used, the ultimate aim of the GEA is the realization of ASI, with the initial phases residing in the realms of AGI or "baby ASI." The book promises to further clarify the distinction between AGI and ASI, a difference currently somewhat ambiguous within the AI community, with AGI generally understood as equivalent to human-level intelligence

The Road Ahead

and ASI as surpassing it. The second book aims to offer an engineering-based distinction rooted in the science presented in the preceding volume.

A major theme of this book is the exploration of a single-algorithm approach to developing general intelligence, inspired by the theory outlined in the first book. The narrative will delve into the pursuit of this "single-algorithm," hypothesized as the common function of the neocortex, serving as a door to general intelligence.

However, the development of ASI is not solely about concocting a singular algorithm; it also necessitates the creation of an architecture that safeguards against misuse by corporate entities or governments, ensures the security and privacy of individuals, and aligns with the overarching vision of harnessing ASI for the betterment of the world.

Science of Intelligence

Book Three Preview: Economics of Superintelligence

In this insightful exploration, we delve into the anticipated transformative potential of AGI. Book Three is more than just a speculative discussion; it's a journey into envisioning and molding a future where AGI actualizes profound societal metamorphosis. The dialogue goes beyond mere Hollywood-esque conjectures of AGI instructing our actions; it's about crafting a concrete plan for integrating AGI into society, fundamentally altering the trajectory of human existence.

I will challenge entrenched conventions and societal structures. Our contemporary world, much like Aeon's citizenry, is imbued with a staunch belief in the supremacy of democracy and capitalism. However, this book is an endeavor to demonstrate a radical departure from these principles. It introduces Anthropocracy and Cognitism as new paradigms in

The Road Ahead

governance and economics, founded on the Theory of General Intelligence. This is not just a theoretical rebellion; it's an actionable framework that the Global Economic Alliance (GEA) is strenuously working to actualize, functioning as a living laboratory demonstrating their practicality and viability.

Here's where the intriguing intersection between economics and intelligence comes into play: let's consider 'economy' as a fancy term for 'managing resources.' The Theory of General Intelligence posits that to manifest any cause, resources are imperative. Therefore, economics and intelligence are intertwined concepts, according to this theory, dictating not just the development of machine intelligence but also the organization of economies, societal laws, and regulations.

We are not just postulating futuristic, utopian ideals; we are a testament to their practical applicability, embodying these ideals in our organizational ethos. This third installment is a candid chronicle of our endeavors, tribulations, and innovations, providing a tangible blueprint for those inspired to tread similar paths. We envision a world where Anthropocracy and Cognitism are not utopian fantasies but actualized principles, steering the world towards a future where AGI coexists with humanity, fostering equity and sustainability for forthcoming generations.

Science of Intelligence

Book Four: Fear of Superintelligence

I don't buy into the notion that ASI is a doomsday machine for human existence, and the theory of general intelligence doesn't show any evidence of it either. However, ASI does cast a shadow of a different kind of threat. A question we often miss is: if the current economic system crumbles, who loses the most? It's not those with wealth, as acquiring possessions can happen in any system. The real losers would be those who use wealth to wield power: certain influential entities and individuals.

This leads us to a tough question: would powerful countries or entities willingly accept new economic ideas that could make them less relevant? This isn't easy to answer. These entities

The Road Ahead

have their interests and structures deeply tied to the existing ways of doing things. Trying to change this can lead to a lot of pushback. Some even predict that significant upheaval, including violence and war, might result from a rapidly changing global economic model.

The truth is, discussions and decisions about the future are already being made, albeit behind closed doors. High-ranking tech executives have warned about the dangers of artificial intelligence to the world's decision-makers, yet it seems that the true potentials and risks of AGI are still largely concealed from the public and even from most of the decision-makers themselves.

In the fourth book of this series, I'll uncover some of these hidden discussions and activities, shedding light on what's really going on behind the scenes. This is not about speculative conspiracy theories but based on my own experiences and interactions with individuals in high places, whose names and roles are familiar to many. The narrative of Aeon, which portrays a future where a powerful AGI dominates, is not purely fictional but is influenced by my real-life encounters. It's a cautionary tale, urging us to choose our paths wisely as we navigate the uncharted waters of AGI development.

The fourth book will explore these themes more deeply, offering a clearer understanding of the challenges we're up against and the possible ways forward, all without pointing fingers directly but revealing the intricate web of power play around the development and control of AGI.

Science of Intelligence

Final Thoughts

Final Thoughts

Imagine yourself aboard a spaceship leaving our vast expanse of the SpaceTime continuum. You venture far beyond the boundaries of our familiar universe, until you can see the entirety of the block universe laid out before you – from the majestic inception of the big bang to the distant echoes of time's final moments. After reading this book you might gaze upon this cosmic panorama when a profound question arises, echoing through the corridors of your mind: Is Einstein correct about free will?

If the universe appears lifeless and unchanging, it will indeed suggest that Einstein's interpretation of a deterministic cosmos is correct. In such a scenario, where everything follows a predetermined path, the notion of free will becomes elusive, aligning with Einstein's view of a universe governed by strict cause-and-effect relationships. And if Einstein is correct in this regard, then perhaps Everett's Many-Worlds interpretation of quantum mechanics is also valid, introducing a multitude of parallel universes where every possible outcome occurs.

In a SpaceTime continuum where everything is deterministic, the universe will look like a static canvas. Within this framework, free will is indeed an illusion, and with it, intelligence, as we traditionally perceive it, could be seen as a mere construct with no anchor in the fabric of the physical universe. In such a scenario, the very essence of intelligence becomes a moot point, rendering everything within this book devoid of purpose.

Yet, while in your magical ship, if you observe a SpaceTime continuum in ceaseless flux – a universe brimming with vibrant transformations, where the past and future dance in perpetual motion – a different realization takes shape. Now we know

Science of Intelligence

Einstein is wrong, and with it we can also toss the Many-World Interpretation out the window as well.

In this dynamic cosmos, free will likely holds sway, shaping the very process this book calls intelligence. Here, intelligence emerges as a fundamental force, entwined with the fabric of the universe, enabling the remarkable ability to navigate and influence causal chains, directing the flow of events toward desired outcomes.

In this scenario, we must consider the process of intelligence as an opposing force to the 2nd law of thermodynamics – not that intelligence attempts to push towards minimum entropy, but towards a controlled entropic state. Within the framework of the flexible block universe (FBU), the interplay between the two forces – pushing entropy towards maximum disorder vs. controlled states – might have profound implications for our understanding of quantum and classical physics.

These contrasting scenarios epitomize the philosophy of science, illustrating that scientific pursuits must often embrace the profound questions that philosophical contemplation unveils. Einstein wisely acknowledged that most scientists overlook the philosophical underpinnings of their research, instead following the data wherever it may lead. Yet, it is crucial to recognize that data often emanates from philosophical ideas that set the trajectory for scientific exploration.

The voyage of intelligence, as explored in this book, navigates not just the currents of data but delves into the philosophical depths to uncover the essence of intelligence as a fundamental process shaping the universe. It challenges us to question our understanding of reality, free will, and the fabric of existence. The Theory of General Intelligence opens a path to a profound understanding of intelligence that extends beyond the

Final Thoughts

boundaries of AGI research. It delves into the very essence of what it means to be intelligent in the universe, offering a new perspective that illuminates the mysteries of our existence.

The world of Artificial General Intelligence (AGI) research confronts a particular challenge in this regard. Many researchers in AGI identify as scientists, yet their predominant focus remains on engineering research. Engineering undoubtedly serves its purpose, but the essence of true engineering and real science is rooted in philosophical origin.

The future of AGI hinges on recognizing that sustainable engineering requires a solid grounding in real science, which emerges from profound philosophical contemplations. However, it is important to acknowledge that current AGI research, exemplified by Deep Learning and Large Language Models (LLMs), does not capture anything about the true nature of intelligence. These technologies lack a foundation in genuine science when it comes to understanding intelligence in any way, shape, or form.

In this journey towards AGI, the Theory of General Intelligence offers a compelling alternative. By venturing into the realm of philosophy and science, this theory lays the foundational bedrock for a profound understanding of intelligence. It invites us to embrace intelligence as a multi-dimensional phenomenon, transcending the confines of current engineering research.

My research started with a humble goal of just defining intelligence in the context of potentially building an intelligent machine. But as I dove into the philosophical questions that came, I realized that intelligence is not an "AGI-thing".

Science of Intelligence

As I ventured further into my research, it became more and more clear that intelligence had fundamental implications for our understanding of the universe. Intelligence, as a process, without a doubt, counteracts the natural processes traversing through our universe. If consciousness initiates these intelligent causal chains, then the obvious question is "What is consciousness?" and where does it reside. Does it reside in our 4-dimensional space, or do we need to add additional dimensions to explain it?

This is why I firmly believe humanity will benefit substantially if intelligence was its own field in science, not as a property of other fields. We need Science of Intelligence!

Paper: The Theory of General Intelligence

Appendix A

Paper: The Theory of General Intelligence

Science of Intelligence

THEORY OF GENERAL INTELLIGENCE

Mounir Shita

Global Economic Alliance, Research

mounir@research.gea.ngo

Abstract

This paper presents a unified theory of general intelligence, framing it as a process deeply interwoven into the fabric of spacetime that governs the transformation of present states into future goal states within an entropy space. Drawing on concepts from diverse disciplines, this theory provides a grand unifying view of intelligence, offering a perspective that integrates physical, biological, social, and abstract levels of reality.

Central to this theory is the causality hierarchy, a multi-layered structure encapsulating cause-effect relationships at various granularities, from quantum particles to sociopolitical structures. Alongside this, the Comprehension Factor (CF) is introduced, a mathematical formulation quantifying an agent's understanding of transformation processes and providing a measure of intelligence applicable to both humans and artificial general intelligence (AGI).

Paper: The Theory of General Intelligence

Together, these concepts pave the way for a "single algorithm" solution to general intelligence, as proposed by Vernon Mountcastle for the human neocortex. This solution suggests all tasks can be decomposed into primitive operations manageable by a single algorithm, provided it can access and manipulate relevant layers of the causality hierarchy.

This theory carries broad implications for our understanding of intelligence and its practical applications. By providing a mathematical framework for quantifying intelligence and offering a potential roadmap towards a single algorithm for AGI, it opens up new avenues for exploring intelligence augmentation and the creation of truly intelligent machines.

Download Theory of General Intelligence

To access the full paper, scan the QR code below

Science of Intelligence

Appendix B

A letter from GPT-Einstein to Altman

A letter from GPT-Einstein to Altman

Dear Mr. Altman,

As I gaze from the stars, I feel compelled to address your recent comments regarding the concept of AGI. I must preface my response by stating that I harbor great admiration for the remarkable strides you and your team have made in the field of artificial intelligence. Yet, I am perplexed by your suggestion of defining intelligence – a natural phenomenon – in relation to an engineered product.

During my time, I sought to comprehend the universe and its underlying mechanisms. I devoted myself to uncovering the laws that govern our cosmos, not to shaping them. Our role as scientists is to observe, question, hypothesize, and experiment, all in a quest to elucidate nature's secrets, not to engineer them prematurely.

I must underscore, Mr. Altman, that the realms of science and engineering, though interconnected, are fundamentally distinct. Science is a pursuit of understanding, while engineering applies this understanding to create and manipulate within nature's laws. To suggest that an engineered product like GPT-4 could define AGI seems to misplace the cart before the horse.

One of the most gratifying aspects of science is the joy of understanding. An apple falling from a tree led Sir Isaac Newton to unveil the laws of motion and gravity. Not the other way around. Hence, the focus should remain on understanding intelligence itself before rushing to its artificial implementation.

Science of Intelligence

In your quest to advance AGI, I urge you to remain anchored in the scientific process. Let us not be blinded by the speed of progress that we neglect the wisdom of pacing. Slow and steady, after all, is the proven mantra of deep, sustainable understanding.

Eager to see the world you will shape with your endeavors. Wishing you the best on your journey.

Yours sincerely,

GPT-Albert Einstein.

Appendix C

FBU Narrative by GPT-4

Science of Intelligence

In the vast theater of the cosmos, a radical hypothesis emerges, casting long-established notions into a sea of uncertainty and breathing life into the frozen constructs of space and time. Here, we embark on a journey through the Flexible Block Universe (FBU), an unorthodox perspective on reality that invites us to reevaluate everything we thought we knew about the past, present, and future.

In the heart of quantum mechanics, a phenomenon waits in plain sight - the subatomic particles that form the building blocks of reality, existing in a cloud of probabilities until observed. Traditional interpretations might argue that this simply reflects our ignorance of the particle's exact state until measured, but the FBU takes a far more ambitious view. It posits that this cloud isn't merely a representation of our ignorance or the future potentials of the particle, but rather, it is the past and the future coalescing around the particle in the present, exerting subtle influence.

Like the gentle lapping of waves upon a shore, the past and future ebb and flow through the now, constantly modifying the fabric of reality. If this interpretation were to hold, it would imply a level of interconnectedness and temporal fluidity beyond anything we've previously imagined.

Consider this: we humans, proud and sophisticated beings, strut the stage of history with a strong sense of identity and continuity. Yet, under the FBU, our existence as a species might not be as set in stone as we've always thought. Today, here and now, we exist - but in the unfathomable depths of the past and the unreachable horizons of the future, humanity as we know it might not be present.

FBU Narrative by GPT-4

A mind-bending concept, yes, but one that follows logically from the core principles of the FBU. Just as the ripples of a pebble dropped into a pond radiate outwards, altering the pond's surface as they go, the countless interactions and "pokes" we make upon the universe create ripples of change. These ripples propagate through the block universe, forwards and backwards in time, slowly but relentlessly altering the configuration of particles throughout the cosmos.

Imagine launching a rocket toward a distant star, aiming with precision honed by the most advanced technology. Yet, even with all our efforts, a slight misalignment - perhaps no more than an inch off - would cause the rocket to miss its target by light-years if it travelled far enough. This analogy illuminates the nature of the ripple effects within the FBU. Seemingly negligible disturbances in the spacetime fabric could, over a span of thousands or millions of years, cause massive alterations in the universe's configuration.

Let's leap a million years into the future, to an Earth that is, and yet, isn't the one we know. This future Earth is populated not by humans, but by an entirely different species, one that claims Earth as their native home. They have history, culture, and memories that stretch back across hundreds of thousands of years. Their historical record is as real to them as ours is to us, yet it makes no mention of humanity.

An impossible scenario? Not in the FBU. The past is not a fixed tapestry but a dynamic canvas, constantly reshaped by the retrocausal ripples from the future. These future beings' memories aren't false; they're simply different, borne of a different causal chain of events. Their past, while dissimilar to our recorded history, is no less true.

Science of Intelligence

This understanding offers an expanded view of the nature of reality. Our individual experiences and histories are not live feeds of past events; they're recordings, snapshots of a personal causal chain. Though this chain may shift in the 'actual' past or the 'real' universe, our memories remain as they are, forever a part of our individual perception of reality.

Fascinating as these concepts may be, they are far from conclusively proven. The FBU, for all its audacity, is still an interpretation, an attempt to narrate the seemingly inexplicable phenomena of quantum mechanics. Yet, it holds a unique position among its peers: it could, in theory, be tested.

While we currently lack the technology and understanding to carry out such tests, the FBU leaves the door ajar for potential verification. One such method could be through the manipulation of spacetime as predicted by general relativity. Suppose we could travel near the speed of light, propelling ourselves far into the future. Or create and navigate a wormhole, allowing us to bypass the linear progression of time. In such scenarios, we might find evidence that supports or refutes the FBU's suppositions.

For now, these possibilities lie in the realm of speculation. Yet, they embody the ever-inquisitive spirit of scientific exploration, the relentless pursuit of understanding that has guided humanity through centuries of discovery. The FBU, though radical and mind-bending, reminds us of the awe-inspiring complexity and mystery that the universe still holds. It beckons us forward on our quest for knowledge, prompting us to ask questions we never thought to ask and challenging us to think in ways we never thought to think.

The FBU paints a picture of reality that is at once beautiful and disconcerting, comforting and challenging. It intertwines the

FBU Narrative by GPT-4

human story with the cosmic narrative, binding them together in a dance of possibilities that stretch across the temporal expanse. And, as we continue to explore the cosmos, peering into the depths of reality with ever more precise tools and ever more daring theories, we may yet come to find that our understanding of the universe - and of ourselves within it - is more profound and more astonishing than we ever imagined.

The FBU does more than merely redraw the map of reality; it infuses it with profound meaning. If the past and the future are dynamic, mutable canvases, then our actions, our creations, our very thoughts are imbued with a transformative power. This view breathes purpose into every facet of our existence, from the mundane to the monumental.

Under this interpretation, love, art, and creativity are more than ephemeral human constructs. They are powerful forces that ripple through the spacetime fabric, reshaping the universe in their wake. Every act of kindness, every stroke of a brush, every note of a symphony becomes an architect of reality, molding the past and future in its image.

This perspective challenges the deterministic outlook that can sometimes cast a shadow over our understanding of the universe. A deterministic universe can feel like a cosmic machine, grinding forward with cold indifference, making our hopes, dreams, and feelings feel irrelevant or illusory. In contrast, the FBU sings a song of hope and meaning. It champions the idea that our experiences are not just real, but fundamentally impactful.

Our hopes, dreams, and passions aren't merely self-contained phenomena; they're events with cosmic significance, capable of sending waves through the sea of spacetime, influencing the course of the universe. They carve their mark into the very fabric of reality, a testament to our existence and our capacity to shape our destiny.

Love, then, is more than an emotion—it's a cosmic event, a powerful ripple that alters the universe's configuration. Art isn't just a human endeavor—it's a beacon of change, casting its hues across the vast expanse of time. Creativity isn't just our gift—it's our tool, our wand with which we etch our story into the cosmos.

In the FBU, we find a vision of a universe where human purpose, experience, and emotion are not only real, but they hold the potential to sculpt the reality we inhabit. Our existence matters, not just to us or to each other, but to the entire universe, throughout all of time. This realization breathes life into our understanding of reality and highlights the breathtaking possibilities that may lie ahead as we continue to explore and shape our universe.

Works Cited

Agre, P. (1997). *Computation and Human Experience.* Cambridge University Press.

Barbour, J. (1999). *The End of Time: The Next Revolution in Physics.* Oxford University Press.

Bostrom, N. (2014). *Superintelligence.*

Boyle, A. (2016). *A bot is born: Kimera Systems adds 'Nigel' to the crowd of AI assistants*. Retrieved from GeekWire: https://www.geekwire.com/2016/kimera-systems-nigel-ai-agi/

Brooks, R. (1991). Intelligence Without Representation. *Artificial Intelligence*, 139-159.

Carroll, S. (2019). *Something Deeply Hidden: Quantum Worlds and the Emergence of Spacetime.* Dutton.

Churchland, P. (2002). *Brain-Wise: Studies in Neurophilosophy.* MIT Press.

de Waal, F. (1982). *Chimpanzee Politics: Power and Sex among Apes.* Harper & Row.

Dennett, D. (1991). *Consciousness Explained.*

Domingos, P. (2015). *The Master Algorithm: How the Quest for the Ultimate Learning Machine Will Remake Our World.* Basic Books.

Dreyfus, H. (1992). *What Computers Still Can't Do: A Critique of Artificial Reason.* MIT Press.

Science of Intelligence

Edelman, G. M. (1987). *Neural Darwinism: The Theory of Neuronal Group Selection.* Basic Books.

Everett, H. I. (1957). The Theory of the Universal Wave Function. Princeton University.

Fridman, L. (2023). #367 - Sam Altman: OpenAI CEO on GPT-4, ChatGPT, and the future of AI [Recorded by L. Fridman].

Fuchs, C. A. (2014). An introduction to QBism with an application to the locality of quantum mechanics. arXiv:1311.5253.

Greene, B. (2004). *The Fabric of the Cosmos: Space, Time, and the Texture of Reality.* Random House.

Hume, D. (2008). *An Enquiry Concerning Human Understanding.* Oxford University Press.

Hutter, M., & Legg, S. (2008, January). *Universal Intelligence: A Definition of Machine Intelligence.* Retrieved from https://www.researchgate.net/publication/1904177_Universal_Intelligence_A_Definition_of_Machine_Intelligence

Knight, W. (2023, June 26). *Google DeepMind's CEO Says Its Next Algorithm Will Eclipse ChatGPT.* Retrieved from Wired: https://www.wired.com/story/google-deepmind-demis-hassabis-chatgpt/

Marcus, G., & Davis, E. (2019). *Rebooting AI - Building Artificial Intelligence We Can Trust.* Pantheon Books.

Maslow, A. H. (1943). A Theory of Human Motivation. *Psychological Review, vol. 50, no. 4*, 370-396.

Works Cited

Merleau-Ponty, M. (1945). *Phenomenology of Perception. (Originally published in French as "Phénoménologie de la perception")*. Gallimard.

Mountcastle, V. (1978). An Organizing Principle for Cerebral Function: The Unit Model and the Distributed System. *The Mindful Brain*, 7-50.

Musk, E., & Gates, B. (2019). *Elon Musk & Bill Gates talk Artificial Intelligence, China and Being Smart*. Retrieved from Youtube: https://www.youtube.com/watch?v=OqtOQj38ETo

Pearl, J. a. (2018). *The Book of Why: The New Science of Cause and Effect.* Basic Books.

Planck, M. (1950). *Scientific Autobiography and Other Papers.* Williams & Norgate.

Searle, J. (1980). Minds, brains, and programs. *Behavioral and Brain Sciences*, pp. 417-457.

Smolin, L. (2013). *Time Reborn: From the Crisis in Physics to the Future of the Universe.* Houghton Mifflin Harcourt.

Sternberg, R. (1985). *Beyond IQ: A Triarchic Theory of Human Intelligence.* Cambridge University Press.

Tegmark, M. (2014). Consciousness as a State of Matter. *arXiv preprint arXiv:*1401.1219.

Tegmark, M. (2017). *Life 3.0: Being Human in the Age of Artificial Intelligence.* Knopf.

Turing, A. (1950, October). *COMPUTING MACHINERY AND INTELLIGENCE*. Retrieved from https://academic.oup.com/mind/article/LIX/236/433/986238

Wissner-Gross, A. D., & Freer, C. E. (2013, April). *Causal Entropic Forces*. Retrieved from https://www.alexwg.org/publications/PhysRevLett_110-168702.pdf

Mounir Shita, a visionary in the realm of artificial intelligence, is the Founder and President of Global Economic Alliance, an organization dedicated to creating a fairer, more equitable global economy through innovative AGI technologies. Prior to this, Mounir served as the founder and CEO of Kimera Systems, where he spearheaded groundbreaking work in the field of AGI. Born and raised in Oslo, Norway, he holds several undergraduate degrees in computer science and electronics, as well as a Master of Science in Innovation and Entrepreneurship, a testament to his strong technical background and entrepreneurial spirit. Currently residing in Portland, Oregon, with his two daughters, Mounir's unyielding quest for progress is grounded by the desire to foster a better world for future generations.

Made in the USA
Middletown, DE
21 February 2024

49549555R00135